T0298358

Design of Structural Steel Joints

This book presents a systematic method of learning how to design perfect joints for steel buildings in industrial projects. It describes the types of joints, details different types of jointing, and covers the mechanics of joints, supported by worked-out examples for different situations. It also includes design charts for full-strength joints of all standard sections. The design and details presented in this book conform to Indian codes and US standards for general building and structural steel work.

Features:

- Provides details on connection design principles and applications from an application point of view.
- Covers practical aspects and good engineering practices related to connection design.
- Explains mechanics of joints with illustrations and sketches.
- Includes design charts for full-strength member joints of standard sections.
- Covers worked-out examples (sketches with supporting calculations) of all typical connections from roofs to base plates.

This book is aimed at professionals in civil and structural engineering, steel structure design, and detailing.

Design of Structural Steel Joints

Ashoke Kumar Dasgupta

CRC Press
Taylor & Francis Group
Boca Raton London New York

CRC Press is an imprint of the
Taylor & Francis Group, an **informa** business

Designed cover image: Ashoke Kumar Dasgupta

First edition published 2025
by CRC Press
2385 NW Executive Center Drive, Suite 320, Boca Raton FL 33431

and by CRC Press
4 Park Square, Milton Park, Abingdon, Oxon, OX14 4RN

CRC Press is an imprint of Taylor & Francis Group, LLC

© 2025 Ashoke Kumar Dasgupta

ISBN: 978-1-032-52794-9 (hbk)
ISBN: 978-1-032-88688-6 (pbk)
ISBN: 978-1-003-53912-4 (ebk)

DOI: 10.1201/9781003539124

Typeset in Times
by SPi Technologies India Pvt Ltd (Straive)

To my parents

Contents

Preface

There has been a long-standing need for an experimental book on the design of structural steel joints. A comprehensive demonstration of steel connection design is often not included in many university curriculums. The structural steel design drawing goes into the fabricator's drawing offices where the workshop drawings showing jointing materials — bolts, welds, and sizes of connection plates — are prepared for fabrication of structural elements. This detailed engineering is generally developed by draftspersons, associate engineers, and mechanical engineers, and many of them do not have adequate exposure to the mechanics of joints. The aim of this book is to provide the necessary tools for them to understand and develop joint design and encourage them to solve atypical problems.

The scope of this book includes basic rules of joint design, methods of strength design of joint materials, and how to use them economically in transferring forces from one joint member to another. Workout design calculations of joints — both welded and bolted, design charts for full strength member joints and typical drawings of joints — are presented for guidance.

This book covers the basic principle and style of work followed in the USA and India. It is my sincere belief that this book will be useful for both students and practitioners in all countries.

Ashoke K. Dasgupta
B.E.(Civil), F.I.E
Chartered Engineer (India)

Acknowledgements

To my colleagues who demanded this book and to my family and publisher who pursued me to write this book. Special thanks to the team of Taylor & Francis; my brother-in-law, Mr. Subhendu Baran Roy; and my son, Dr. Suman Dasgupta, for suggesting corrections and improving the manuscript.

About the Author

Ashoke Kumar Dasgupta has worked in the prestigious consulting engineering companies Development Consultants Pvt Ltd (India) and The Kuljian Corporation (USA) at their design offices and construction sites in India, USA, and Bangladesh; Middle Eastern countries: Turkey, Saudi Arabia, Syria, Egypt, UAE, Qatar, and Kuwait; and Far Eastern countries: China, Japan, Indonesia, and Vietnam from 1974 to 2019.

His expertise lies in Structural, Civil, and Foundation engineering for Power Plants and Industrial Structures.

This book is his second publication with Taylor & Francis. The previous book was *Design of Industrial Structures: Reinforced Cement Concrete and Steel*, published in 2022.

1 General

1.1 INTRODUCTION

The performance of a building or structure depends on the efficiency of its member end connection details.

The details of connections should be simple and easy to approach. The designer should select appropriate types of joints for effective transfer of forces and the conditions of restraint or movement assumed in analysis of the building framework, i.e., simply supported, hinged, or fixed joint. The shape of connecting gusset plate and length of weld or the array of bolt group shall be designed to ensure effective transfer of member forces without developing concentration of local or secondary stress as far as possible.

In practice, the connections are designed for the full strength of the member section unless the member forces are given in design drawing.

Nature is the best reference for understanding what the appropriate style of joint design and detailing should be. The joints between the trunk and branches of a tree or the joints in our human body that allow movement in different directions without stress concentration are natural examples of joint design.

1.2 TYPES OF JOINTS AND CONNECTION ELEMENTS

The types of joints commonly used for general building and structures are as follows:

a) Shear connection
b) Moment connection
c) Sliding connection

Shear connection is considered as a simple connection, which is used to transfer the end REACTION of a beam or girder to a supporting column. The shear connection must be able to meet the required rotation and must not introduce strength and stiffness that significantly alter the rotational response. In analysis, this type of connection is defined as pinned of hinge connection. The shear connection is provided by one or two web cleats or plates joining the web of the supported beam and face of the supporting column or beam or girder. Chapter 2 shows details of a shear connection joint.

Moment connection joints are fully restrained joints that have sufficient strength and stiffness to transfer the end moment and the shear to columns, maintaining the angle between beam and column. The Moment Connection may be idealized as a fixed joint in the structural analysis. Chapter 2 includes details of moment connection joints – welded and bolted.

Partially restrained moment connections are designed to transfer moment but also allow rotation between connected members to satisfy design assumption. In this type

DOI: 10.1201/9781003539124-1

of connection, the connection elements must have sufficient strength, stiffness, and deformation capacity within allowable limits. When a partial restrained connection is used, the relevant response characteristics of the connection must be included in the analysis to determine the member and connection forces, displacement and the frame stability.

Sliding connections are provided to allow lateral or vertical movement of joints, per assumptions considered in structural analysis. These connections connect walk-way or gallery structure, conveyor supporting structure, crane girder, expansion joints and bridges in order to allow lateral movement due to thermal stress, differential settlement, etc. There are various types of sliding joints. Slotted holes in connection cleats allow the slip or movement of supported members at joints. Sliding surface is also used at the bearing point by providing stainless steel liner over the sole plate or Teflon-coated steel plate to ease movement of the member at support. The gable end columns at buildings or shed-type structures are also seen to have slotted cleats or plate at column cap, which allow vertical deflection of roof structure. Chapter 6 shows examples of sliding connection.

Hinged connections are used at conveyor gallery supports and bridge bearings. A hinge connection transfers vertical and horizontal reaction only. It allows rotational movement with no restraint and cannot transfer end moment. A perfect hinge connection joint is expensive. In building and pipe rack structure, relatively simple types of joints are detailed with seating bracket and web cleats. Such joints allow release of moment but with partial fixity, which is ignored for all practical purposes.

A joint requires **connection elements** such as angle cleats, gusset plates, brackets, and connectors – bolts and welds.

Bolted connection
  i) Bearing type bolt
  ii) Friction grip – slip resistant type bolt

Welded connection
  i)   Fillet weld
  ii) Butt weld
  iii) Stud weld

In general, bolted connections are used for field joints, and welding is used for fabricated joints at workshop. Field joints can also be welded, per designer's choice. The welding at site should be done under controlled supervision and tests performed in conformity with standard codes and manuals.

Nominal diameter of bolts shall not be less than 16 mm (M16), and minimum size of fillet weld shall not be less than 6 mm for permanent connections.

Combinations of two types, i.e., bolt and weld, should not be done at the same joint. When using different fastening material, for example – (i) connecting flange plates by welding or (ii) connecting webs with cover plates and friction grip bolts – the compatibility of the deformation must be checked. However, these types of mixed jointing procedures should be avoided.

TABLE 1.1
Specifications of Materials

SL No	Joint Material	ASTM Standards	Indian Standards	British Standards
1	Angle cleats, plates, tee, cut joist and channels, chequered plate, gratings and fabricated components	A36 (Fy = 36 ksi)	E 250 (Fe 410W) A – IS 2062	S275 BS EN 10025
2	Bolt and nut	A325, A490	IS 1363, IS 1364, IS 1367, IS 3640, IS 3757, IS 4000, IS 6623, & IS 6639	HSFG – General Grade BS 4395-1 & BS 5950-2
3	Washers	F436 Type 1	IS 5369, IS 5370, IS 5372, IS 5374, IS 6610, & IS 6649	Compatible to grade of bolt
4	Weld consumable	AWS D 1.1	IS 814, IS 1395	Class 35 BS 5950
5	Foundation anchor bolt	A307 or A193	IS 5624	BS 7419

1.3   MATERIALS USED FOR JOINTS

The material of components shall conform to the American standards or equivalent international standards accepted by the project authority. The standard specifications are given above for the reader's guidance (Table 1.1).

1.4   RULES OF CONNECTION, DRAWINGS AND STANDARD NOTES

1.4.1   RULES OF CONNECTION

The basic rules of connections are as follows:

1. The center of gravity of the weld group (weld seam group) or bolt group shall coincide with the gravity axis of member parts to be connected. If there is an offset, the connector weld or bolt group shall be designed for the moment due to eccentricity.

   For truss member joints, designed on the assumptions of pin jointed connection, members meeting at a joint should, wherever practicable, have their centroidal axes meeting at a point – and the center of resistance of a connection should lie on the line of action of the load so as to avoid an eccentricity moment on the connections. However, where the eccentricity of members or of connection is present, the members and the connections shall provide adequate resistance to the induced bending moments.

2. The individual cross-sectional parts, e.g., flange, webs in a splice joint of beam section, shall be joined individually in accordance with the proportionate forces being carried by the individual parts. If the connection is made in part, e.g., only webs are joined at splice location, the deflection of the member at that splice location shall be checked as per permissible limit.

3. Where a connection is subject to impact or vibration or to reversal of stress (unless such reversal is solely due to wind), or where for some special reason – such as continuity in rigid framing or precision in alignment of machinery – slipping of bolts is not permissible, high-strength friction grip bolts (non-slip) or welding shall be used. In all other cases, bolts in clearance hole (bearing type) may be used. Snug-tightened high-strength bolts are recommended for bearing connections.

4. The connections shall be designed consistent with the end conditions (fixed, hinged or sliding) assumed in the analysis of structure.

5. Ease of fabrication and erection should be considered in design of connections. All fastening bolts and nuts, weld seams should be accessible for tightening, welding, inspection and painting at site. All members should have necessary erection clearance at ends.

6. The minimum and maximum spacing of bolts and minimum edge and end distance of bolt holes from the side and end of a member shall be within limit appropriate code and standard. However, the minimum spacing of bolts shall not be less than 2.5 times diameter of the bolt and the minimum edge or end distance not less than 1.5 times diameter of the bolt hole.

7. Length of cover plates used at splice location shall be adequate to receive stress satisfying moment requirements.

8. Diameter of bolts shall not vary on the same joint.

9. The stresses of all components – including bolts, nuts and weld – should be kept within limits specified in American/British/Indian or other international standard, as applicable for the project site.

1.4.2 DRAWINGS AND STANDARD NOTES

Drawing is the language of engineers learned at the beginning of engineering courses. The structural steel **design drawings** are prepared to outline framing structure in plan, elevations, sectional views and description of member size and type, e.g., rolled steel section or built-up item. It shows grid line dimensions, story heights and details of major connections of members. The **fabrication drawing or workshop drawing** shows enlarged details of each member including end connections and provisions for joining other members to it so that they fit together. This drawing shows (i) cutting length of primary members, (ii) gussets or cleats attached for connecting other members as per design drawing, (iii) location and size on connecting holes for bolts and weld, and (iv) actual shape and dimension of each component with respect to working point (where the centroidal axes meet at a point). A list of materials generally called **Bill of Materials** are presented in this fabrication drawing, which covers weight and size of all parts and members. An example of the bill of materials is given in Table 1.2.

TABLE 1.2
Bill of Materials

Erection Mark	Part No.	Item No.	Item Description	Length (mm)	Nos Required	Unit Weight (Kg/m or Kg/sqm)	Weight of Item (Kg)	Weight of Each Erection Marked Item	Number of Each Erection Marked Item Required	Total Weight (Kgs)	Remarks
G/CBR-1 &G/	1		L 150 × 150 × 16	5390	2	35.80	386				
CBR-1X	2		PL 115 × 10	170	7	78.50	10.74	397	1 + 1	793	
								Grand total =		793	Kg

Note: L: Rolled steel angle section; PL: Plate.

At fabrication shop, this drawing is used for layout, cutting, and fabricating the individual members. Each item is separately marked in the marking plan drawing, which is a general arrangement drawing that includes the framing plan and unique item number of each member. These item marks and their orientation (North-South face) are painted on fabricated members. The working persons at the site use this drawing to place the columns on the foundation pedestal, erect connecting beams and bracings, and thus complete the framework, as per the drawing. The members are held in position by erection bolts and temporary guy ropes until the final alignment, grouting, and bolting. After the leveling and alignment is complete, the erection bolts are replaced by permanent bolts or welding in conformity with the fabrication drawing.

TEKLA XSTEEL software (Microsoft Windows) is used to prepare fabrication drawing- arrangement and connection detail. The digital information from this software can be directly used by machines at the workshop to fabricate items.

The material specifications, standards and codes are provided in the standard notes of design and fabrication drawings. The **Standard Notes** given below are for guidance to the reader.

A. General

1. All dimensions are in millimetres except floor elevations and plant grids, which are in meters unless stated otherwise.
2. All elevations are with respect to plan building ground floor finish level as EL. (+/−) 0.00 M., which corresponds to RL....... Reduced Level with respect O.S Datum or Mean Sea Level).
3. All drawings shall be read in conjunction with the terms, conditions, and specifications of the contract.
4. The fabrication and erection of structural steel shall be in accordance with American/British/other equivalent international standards, codes and specifications for general construction in building and steelwork.

B. Structural Material

5. All materials shall be new, clean, rust-free, and straightened, if necessary, before being used.
6. Following material shall be used:

   a) All structural steel consisting of rolled steel joists, Tee, channels, angles, pipes, plates and flats shall be of grade...... (yield stress) conforming to codes ......*ASTM/BS/IS/others (Refer Table 1.1)*.
   b) All structural steel for gratings and chequered plate shall be of grade ..... (yield stress) conforming to codes ... *(Refer Table 1.1)*.
   c) All anchor bolts shall be of grade..... conforming to codes ... *(Refer Table 1.1)*.
   d) All connection bolts shall generally be high-strength steel hexagonal type of grade *(8.8)* conforming to codes ... *(Refer Table 1.1)*.
   e) All nuts shall be hexagonal type compatible to grade of bolts.

f) All flat, circular and hardened steel washers shall be compatible to grade of bolts.

g) All welding electrodes shall be of class.......... conforming to code *(Refer Table 1.1)*.

C. Connections

7. All shop connections shall be welded and field connections shall be bolted using high-strength bolts (Bearing type connections) unless stated otherwise.

8. Minimum diameter of bolts shall be 22 mm for primary members and 16 mm for secondary members.

9. The size of fillet weld shall be 6 mm minimum unless noted otherwise.

10. All connections shall be developed in line with good engineering practice and subject to the approval of the engineer.

11. Unless otherwise stated in the design drawings, the connections shall be designed for the following loads:

    a) Moment connections designated as "MC" to be designed for the full capacity of the member.

    b) Shear connection for beams to be designed for half the maximum uniform allowable load (for beams laterally supported) or 70% of the shear capacity of the member, whichever is more.

    c) Connections for the beams in column line to be designed for the minimum axial load of 75 kN (tension/compression) unless higher load is specified, in addition to required shear/moment load.

    d) Column splices to be designed for full capacity of lower sections.

    e) Connection for bracing member to be designed for full strength of the member.

12. All bolt connections must have minimum two bolts.

13. Length of fillet welds shall not be less than four times the nominal size of the weld or 40 mm, whichever is more.

14. Fillet weld terminating at ends must return around corners not less than twice the nominal size of the weld.

15. Butt welds shall completely penetrate unless stated otherwise.

16. The diameter of bolt holes should not be more than 1.5 mm (1/16 inch) than that of the bolt diameter unless noted otherwise.

17. The erection clearance for end of members shall not be more than 2 mm at each end unless stated otherwise.

18. Gusset/End plates shall be of following minimum thickness unless specified otherwise:

    a) Bracings, with member thickness 8 mm or less: 8 mm

    b) Bracings, with member thickness more than 10 mm: 12 mm

    c) All other members: 10 mm

19. All Handrails, Gratings and Chequered plates shall be galvanized unless stated otherwise.

D. Painting

For Steel Structure (Indoor and Outdoor)

a) Surface Preparation: Sand Blast Cleaning (SA 2 – ½ Standards).
b) Primer Coat: One coat of Inorganic Zinc, 75 Micron DFT (dry film thickness).
c) Intermediate Coat: One coat Polyamide Epoxy, 100 micron DFT.
d) Final Coat: One coat Aliphatic Polyurethane, 40 micron DFT.

(Note: The above painting standard was followed in a power plant project in Ireland; however, the designer should select the site-specific coating system as per desired protection life, atmospheric condition and environment classification conforming to relevant codes and specification followed in the relevant country).

1.5   FABRICATION, ERECTION AND TESTING

Fabrication of joints is an important part of member fabrication. The connecting gusset plates and cleat angles with drilled holes are joined with the primary member by welding, following the dimensions and alignment given in workshop drawing.

Mismatching of holes can be avoided if fabrication of the connecting parts is done accurately by the skilled work person at the shop. The end of column shafts or girders, which are part of splice joints, are assembled at the fabrication shop and then drilled together for better alignment of holes. The matching of holes at the erection site becomes easier through this process of drilling a hole through multiple components stacked together.

The members are erected at the site following marking plan drawing. The bolt holes in the end plates or cleat angles of connecting items are matched together and then fitted with erection bolts. After completion of the framework, the structure is aligned and leveled. The erection bolts are then replaced with permanent bolts or welding, as per drawing requirement. Once the joints are complete, inspection and testing of joints is done to ensure safety of structure. Tightening of bolts are checked using calibrated torque wrenches. The field-testing of welds are performed by visual, ultrasonic testing, or X-ray, as per quality assurance procedure indicated in drawing and specifications.

Some important considerations and the best-known engineering methods related to design and detailing of joints are presented below:

a. **Cutting Edges**: Cutting of steel shall be done by shearing, cropping, or sawing. Flame cutting may also be permitted, leaving margin for machining the cutting edge, so that all metal hardened by flame is removed. At least 3 mm shall be deducted from each cut edge to determine effective size of members from gas-cut edge.

b. **Erection Clearance**: The erection clearance for the cleated end of a member meeting at the connecting steel surface shall preferably be not more than 2 mm at each end. The erection clearance at the ends of beams with face plate should not be greater than 3 mm. If more clearance is necessary for practical reasons, the end shear should be transmitted by suitably designed brackets or seating stool.

c. **Bolt Holes**: Holes through more than one thickness of material of members, such as columns and girder flanges at splice joints, must be drilled after the members are assembled and tightly clamped or bolted together. Holes for bearing type or friction grip bolts should not be more than 1.5 mm for bolts up to 22 mm nominal diameter. For bolts having diameters more than or equal to 25 mm, the diameter of the hole shall be 2 mm more than the nominal diameter of bolt passing through them.

d. **Welded Joint**: The parts to be fillet welded should be in close contact as practicable but never be separated by more than 4 mm. If the separation is 1.5 mm or greater, the size of fillet welds should be increased proportionate to the amount of the separation. The fitting of joints at contact surfaces shall be close enough to eliminate water seepage after painting. Abutting parts of the butt-weld joints shall be carefully aligned. Misalignments greater than 3 mm shall be corrected, and in making the correction, the parts shall not be drawn into a sharper slope than $2°$. The job shall be positioned for flat welding (down hand) as far as practicable.

e. **Column Bases**: For columns with a slab base, where the end of the column shaft is designed for complete bearing, the top of the slab base should be accurately machined over the bearing surface for effective contact with the end of the column shaft. Columns with slab bases need not be provided with gussets, but sufficient fastening should be provided to retain the parts securely in place and to resist all the forces and moments, other than direct compressions, including those arising from transient loads. For columns with the ends of the column shaft and the gusset plates that are not faced for complete bearing, the welding connecting them to the base plate should be sufficient to transfer all the forces to which the base is subjected.

Grout holes are provided, where necessary, in the column base plate for the release of air.

f. **Column Splice Joint**: The column splices and butt-joints of compression members depending on contact for stress transmission shall be accurately machined and close-butted over the whole section with a clearance not exceeding 0.1 mm locally at any plain.

g. **Field Bolting**: Friction grip connections are used where slipping of bolts is not permissible in cases such as continuity in rigid framing, precision in alignment of machinery, or connections subject to impact. In all other cases, bolts in clearance holes (bearing type connection) may be used, provided that due allowance is provided for any slippage. Contact surfaces (faying surfaces) within friction type joints should be free of oil, paint lacquer, or galvanizing. There are two types of surface preparation defined in AISC Manual – Class A surfaces (unpainted clean mill scale steel surfaces or surfaces with Class A coatings on blast-cleaned steel or hot-dipped galvanized and roughened surfaces) and Class B surfaces (unpainted blast-cleaned steel surfaces or surfaces with Class B coatings on blast-cleaned steel).

High tensile strength bolts are used in both the types of surfaces mentioned above. When all fasteners in the joint are tight, all high tensile bolts should be tightened to provide the required minimum bolt tension by any of the following methods:

TABLE 1.3
Final Nut Rotations in Turn-of-nut Method

Bolt length not exceeding 8×diameter or 200 mm	Bolt length exceeding 8×diameter or 200 mm	Remark
1/2 turn	2/3 turn	Nut rotation is relative to bolt regardless of the element (nut or bolt) being turned. Tolerance on rotation – 30^0 over or under.

(i) **Turn-of-nut Method**

In this method, enough bolts are required in "Snug tight" condition to ensure that the parts of the joint are brought into good contact with each other. "Snug tight" may be defined as the tightness attained by a few impacts of an impact wrench or the full effort of a person using an ordinary spud wrench. Following this initial operation, bolts should be placed in any of the remaining holes in the connection and brought to snug tightness. All bolts in the joint should then be tightened additionally by the applicable amount of nut rotation specified in Table 1.3, with tightening progressing systematically from the most rigid part of the joint to its free edges. During this operation, there must not be any rotation of the part not turned by the wrench.

(ii) **Torque Wrench Tightening**

When torque wrenches are used to provide the bolt tensions, the bolts shall be tightened to the torque specified in Table 1.4. Nuts shall be in tightening motion when torque is measured. When using torque wrenches to install several bolts in a single joint, the wrench shall be returned to "touch up" bolts previously tightened – which may have been loosened by the tightening of subsequent bolts – until all are tightened to the required torque values. Minimum Bolt Pretension is given in Table 1.5.

h. **Field Welding**: The electrodes to be used should conform to appropriate codes and standard. The covered electrodes are used for manual welding – fillet and butt-weld. Electrodes with hydrogen-controlled (low diffusible

TABLE 1.4
Torque to be applied on High Tensile Bolts
(Unless recommended otherwise in manufacturer's specification)

Nominal Bolt Diameter (mm)	Torque to be Applied (Kg M)
20	56.80
22	77.44
25	102.50

Note: Unless recommended otherwise in manufacturer's specification.

TABLE 1.5
Minimum Bolt Pretension, kN*

Bolt Size, mm	A 325 M Bolts	A 490 M Bolts
M16	91	114
M20	142	179
M22	176	221
M24	205	257
M27	267	334
M30	326	408
M36	475	595

Source: Reference: Table J 3.1M; AISC Manual – 14th Edition.
Note:
* Equal to 0.7 times the minimum tensile strength of bolts, rounded off to nearest kN, as specified in ASTM specifications for A325M and A490M bolts with UNC threads.

hydrogen content) coverings may be used in case of heavy sections being welded. However, these should not be taken as an alternative to preheating of parent metals. Preheating of thick parent metals might be necessary to ensure complete fusion of welding runs (particularly, the initial run) to minimize uneven internal stresses, undue hardening, and distortion. This is also necessary when applying small weld with little heat input to material with thickness, depending on weld pass size, material thickness, carbon equivalent, and type of electrode to avoid hydrogen cracking. While repairing welded joints with small spot welds in thick work pieces, these pieces may need preheating to avoid rapid heat dissipation. For detail of edge preparation and root gap of butt welding, refer AISC Steel Construction Manual.

1.6 JOINTING IN STAIRCASE, LADDER, HANDRAIL, GRATING, AND CHEQUERED PLATE FLOORING

Staircase steps are supported on stringer members. The size of the stringer member should preferably be of 250 mm (10 inch) depth channel section. This is to accommodate standard steps of 250 mm width. The steps should have seating cleat at both ends. The stringer members are connected to supporting beam or column by bolted or welded joint.

Ladders with more than a 2-meter height should be provided with a protective cage. The tall ladders should have a break at every 6-meter interval by providing resting platforms, unless there is a space constraint.

Handrail posts are connected to steel members by bolting or welding. The number of bolts should be a minimum of two.

Grating floors resting on steel supports are fixed by clamps and bolts as per manufacturer's drawing. Intermediate welding may also be used for a fixed type grating floor.

Floors and platforms with **chequered plates** are generally welded to the supporting beam. The chequered plates are designed with or without ribs, depending on load and span.

Standard details are provided in Chapter 6.

1.7 JOINTING HOLLOW SECTIONS

Hollow sections are tubes and pipes made with thin walled steel sections. The field connections are bolted and shop connections are welded. The reader is encouraged to refer to AISC Steel Construction Manual for hollow steel connection design and detailing.

1.8 JOINTING THIN COLD-FORMED SECTIONS – METAL DECK SHEET, METAL CLADDING ON ROOF AND SIDES, PURLIN AND SIDE GIRT

Cold-formed steel (0.8 – 1.2 mm) formed to a shape, like corrugated roof sheets with deep valleys, have many applications. These sheets are used for roofing and side cladding in structural steel building for workshops, storages, and office buildings, as well as in permanent formwork in casting reinforced concrete construction slabs. These sheets are joined side by side by metal stitches. Self-drilling type steel screws are used for fixing on roof purlin, side girt, or supporting steel members. Cold form steels (1.4–3 mm) are also used to prepare Channel Sections (C-Section) and Z-Sections, which are widely used in construction.

The standard details of roof purlin and side girt sections are provided in Chapter 6.

1.9 ANCHOR FASTENER AND EMBEDDED PLATES FOR CONNECTION TO CONCRETE SURFACE

The embedded plates are fabricated with steel plates lugs (flat or bar) welded to it. These items are also named as insert plates, which are used for joining steel members with concrete structure. The embedded plate is fixed in position and leveled before placement of concrete. The top of the plate is leveled and aligned with the finished surface of the concrete member. The lugs are tack welded to reinforcement steel to keep the embedded plate in position during casting. Typical details of embedded plate in concrete are shown in Chapter 6.

Uses of anchor bolts and anchor fasteners are commonly known to engineers. Anchor fasteners are similar to anchor bolts while transferring shear and tension from a plate or fixture attached to a concrete surface. There are two types of anchor fastener generally used in industrial practice – Mechanical anchor and Chemical grout anchor. In the mechanical anchoring process, a small round type (brass or alloyed metal) holding fastener with thread inside and split end is drilled into the concrete surface. During insertion, the split ends expand and become anchored into concrete. The matching bolt is then fitted inside the holding fastener to attach the face plate or fixture on the surface of the concrete. The chemical anchor is an epoxy resin bonded fixture. In this process, holes are drilled into the concrete surface and filled with epoxy resin capsule; the holding fastener is then inserted. During the process, chemical grout – which fills the space around the fastener – becomes solidified inside the hole to generate the required bond. The bolt and fixture are then fitted after the setting time is completed.

1.10   CODES AND STANDARDS

The following Codes and Standards are used for design and construction. The list of codes and standards is not exhaustive but covers a major portion of structural work for general Building and Industrial structures.

American Standards:
- AISC, American Institute of Steel Construction, Manual of Steel Construction
- AWS D1.1, American Welding Society Structural Welding Code for Steel
- ASTM, American Society for Testing and Materials Standards (as applicable)
- SSPC, Steel Structures Painting Council for Protective Coatings
- Uniform Building Code 1997, (UBC 1997)
- Construction Industry Research and Information Association (CIRIA) C577, Guide to construction of reinforced concrete in the Arabian Peninsula
- ACI 318, American Concrete Institute, Building Code Requirements for Structural Concrete and Commentary
- AISC Steel Design Guide – 1 Base Plate and Anchor Rod Design

Indian Standards:
- IS: 800 General Construction in Steel – Code of Practice
- IS: 1367 Technical Supply Conditions for Threaded Steel Fasteners (all parts)
- IS: 3757 Specification for High Strength Structural Bolts
- IS: 9595 Metal Arc Welding of Carbon and Carbon Manganese Steels-Recommendations
- IS: 2062 Steel for General Structural Purposes – Specification
- IS: 806 Code of Practice for Use of Steel Tubes in General Building Construction
- IS: 4000 High Strength Bolts in Steel Structures – Code of Practice
- IS: 4923 Hollow Steel Section for Structural Use – Specification
- IS: 456 Plain and Reinforce Concrete – Code of Practice
- SP 6 – 3: Handbook for Steel Columns and Struts

British Standards and Euro Code:
- Euro Code 3: Design of Steel Structures – Part 1- 8: Design of Joints
- British Standard NA + A1: 2014 to BS EN 1933 – 1: 2005 + A1: 2014
- BS 449-2 Specification for the Use of Structural Steel in Building

- BS 4-1 for Structural Steel Sections
- BS 5950: 2000 (Part 1) Structural Use of Steelwork in Building – Code of Practice for Design – Rolled and Welded Sections
- BS 5427: 1996 (Part 1) The Use of Profiled Sheet for Roof and Wall Cladding on Buildings – Design
- BS 5950: 1994 (Part 4) Structural Use of Steelwork in Building – Design Composite Slab with Profiled Steel Sheeting
- BS 5950: 1998 (Part 5) Structural Use of Steelwork in Building – Design of Cold-formed Thin Gauge Sections
- British Standard, BS 8110 Part 1 & Part 2 for Structural Use of Concrete

REFERENCES

1. Steel Construction Manual AISC – Fourteenth edition.
2. IS 800: 2007 General Construction in Steel – Code of Practice.

2 Joints in Steel Structure

In this chapter, we will also describe the details of some bolted and welded joints that have been used in industrial plant building structures and pipe racks in India, USA, Saudi Arabia, Qatar, Ireland, Vietnam, and other countries. Many types of joint details are followed in structural engineering practice – the details provided here are found as the most popular, easy for erection, safe, and economic.

2.1 BOLTED JOINTS

Two types of bolted connections are used in joint design – Bearing type and Friction Grip or Non-slip type connection. In *bearing type connection*, bolts are subjected to stresses perpendicular to their axes. So, the bearing resistance of bolts on the contact surface at bolt holes and shear strength of bolts across its cross-section should be considered in joint design calculation. In the bearing type joint, the bolts are generally installed in snug-tight condition. The snug-tight condition is defined as the tightness bringing the connected plies into firm contact.

In *friction grip* or *slip critical joints*, the bolts are tightened to a specified torque to generate a *pre-tension force*. This pre-tension force brings the connected plies into firm contact and generates a clamping force on the contact surfaces (faying surfaces) to prevent slip. All high strength bolts to be used in slip critical joints should be tightened to a bolt tension not less than that given in Table J3.1 in kips or J3.1M in kN of AISC manual 14 Ed.

High tensile bolts shall be used for bolting in both the types stated above. Finished holes shall be not more than 1.5 mm or 2.0 mm (as the case may be, depending on the nominal diameter of bolt) in diameter larger than the diameter of the bolt – refer to AISC or applicable code.

Washers shall be used to give the nut and the head of the bolt a satisfactory bearing. The threaded portion of each bolt shall project through the nut at least one thread.

The *erection clearance* for cleated ends of members should preferably not be greater than 2.0 mm at each end. The erection clearance at ends of beams with a face plate should not be more than 3 mm at each end, but where–for practicable reasons–greater clearance is necessary, a suitably designed seating cleat or shear bar or bracket should be provided.

2.1.1 Bolt Value or Design Strength of Bolt

The *bolt value* or *design strength of a bolt* is the capacity of the bolt, i.e., Tension Capacity, Shear Strength, and Bearing Resistance. The strength of a bolt, when subject to tensile load, is estimated by multiplying the net cross section area by yield stress and appropriate factors of safety, as stated in codes of practices in steel design. In joints of shear connection, the capacity of a bolt is determined by its shear capacity

DOI: 10.1201/9781003539124-2

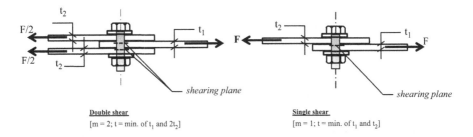

FIGURE 2.1   Bolt value.

Notes: Bolt value will be the minimum of the following:

1. Single shear or double shear of bolt
2. Bearing stress on bolt
3. Bearing stress on plate

and bearing strength, whichever is the minimum. The shear capacity of a bolt is the permissible shear stress of bolt material multiplied by the applicable cross section of bolt shaft and number of shear planes active in the joint. Similarly, the bearing resistance of a bolt in a joint is the product of the contact area in the bolt-hole and permissible bearing stress as per codes. The contact area is the effective plate thickness and nominal diameter of bolt. The bolt value is dependent on the type of joints, numbers, and the thickness of connecting plates in that joint; thus, it should be calculated separately for each group of joint (Figure 2.1).

The above figure shows typical examples of single-bolt-joint for understanding the shear capacity and bearing resistance. The sketch on the right (Single shear) has two plates held together by a single bolt. This joint is subject to a horizontal force, F, acting along the plates moving in the opposite direction. The applied force F is perpendicular to the bolt shank. So, the connecting bolt will experience bearing stress on the contact surface in the bolt-hole and shearing stress across the cross section of bolt. The plane of shear failure of the bolt coincides with the plane of contact between plates.

The sketch on the left (Double shear) shows three plated connection with a single bolt. In this case, the bolt will have bearing support on plate t_1 or $2 \times t_2$. There will be two planes of shear failure along the top and bottom faces of plate t_1. The mathematical expression of bolt value calculation is given below for reference.

a) **For Bearing Type Connection**
   Shear capacity of bolt,

$$Fs = As \times m \times ts \ kN$$

   where,
   As = Cross sectional area of bolt in mm^2.
   m = Number of shear plane (1 for Single shear and 2 for Double shear).
   ts = Allowable shear stress of bolt material in MPa.

Bearing capacity of hole,

$$Fb = s_1 \times d \times t$$

s_1 = Allowable bearing stress on contact surface between hole and bolt shank in MPa (minimum of permissible bearing stress of plate material and bolt material).

d = Nominal diameter of bolt in mm.

t = smallest sum of plate thickness (hole bearing area) in mm.

Now, the Bolt value, Q will be the smallest of Fs and Fb.

$$\text{Number of bolts} = F \div Q$$

b) For Slip Critical Connection

Design strength of bolt will be the frictional force between connected plates developed by the pre-tension in bolt corresponding to applied tightening torque.

In slip critical connection, Bolt value,

$$Q = m \times Fv \div vg \ (kN)$$

μ = Coefficient of friction between contact surfaces = 0.5 (to provide as per code)

Fv = Pre-tension in the bolt (kN)

vg = Coefficient of resistance to slip (1.1 at service load, 1.25 at ultimate load)

Number of bolts = $F \div Q$

2.1.2   BEAM TO BEAM SHEAR CONNECTION WITH CLEAT ANGLES

In this chapter, we will discuss the behavior of joint components and how they are stressed while sharing load between each other. A designer should understand this mechanism before proceeding to the strength design of components in a joint.

Figure 2.2 describes a typical shear connection joint between two beams. The connection components are rolled steel angle cleats, bolts with nut, and washer. The supported beam will transfer, the end reaction ($F = w. L/2$) to the web of the supporting beam through bolts and connector cleats in steps.

In Step I, we will check the web of the supported beam. The web of supported beam has drilled holes for bolt connection. The thickness of the web plate should be adequate to bear the end reaction and transfer the same to the connector bolts. Therefore, we should check the block shear capacity, bearing strength at holes, and local buckling of the web at connection point.

Web buckling – for rolled section, the web thickness is found within the limiting depth-to thickness-ratio specified in codes; hence a local buckling strength check is not necessary. With the built-up girder having adequate web thickness and the

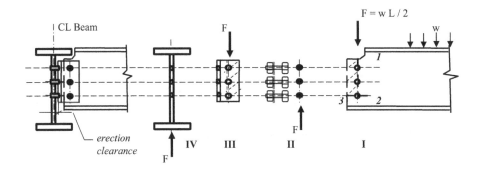

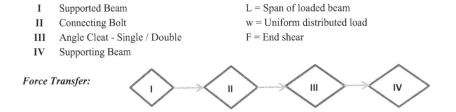

I	Supported Beam	L = Span of loaded beam	
II	Connecting Bolt	w = Uniform distributed load	
III	Angle Cleat - Single / Double	F = End shear	
IV	Supporting Beam		

FIGURE 2.2   Shear connection – with cleat angle.

depth-to-thickness ratio lower than the allowable limit for compact section, the web buckling check is not needed. In case of built-up sections where the depth-to-thickness ratio exceeds allowable limit, the web plate should be reinforced by adding plates or web stiffener to avoid local buckling due to compressive stress developed by end shear (F). The limiting width-to-depth ratio is provided in Table 2 (IS: 800) and Table B4.1b (AISC 360-2010).

The *block shear capacity* of the web at the connection point is the net shear of the web plate along the bolt line parallel to the force (1–2), plus the tension capacity of the plate perpendicular to the line of force (2–3). The hatched part marked in the web of beam shows the lines of probable block shear failure. The *bearing strength* of web at the bolt-hole is the sum of the net projected area of contact (nominal diameter of bolt × web thickness) multiplied by permissible bearing stress.

In Step II, the force F is distributed amongst the bolts in a vertical row. The diameter and number of bolts are determined according to the capacity calculation of individual bolts. The capacity or strength of each bolt is also called *bolt value*. In a bearing type connection, the bolt value will be the shear capacity or bearing strength, whichever is lower. The worked-out example of bolted joints is given in Chapter 4.

In Step III, the end shear F is carried over from bolts to the web cleat (single or double angle). The total thickness of the cleat angle should provide the bearing strength required to support the bolts. The length of web cleat is equal to the sum of the spacing of bolt holes (2.5 to 3 times of nominal diameter of bolt) and minimum edge distances. The cleat is further checked for block shear. Finally, in Step IV, the force is transferred to the supporting beam. The web of the supporting beam is also

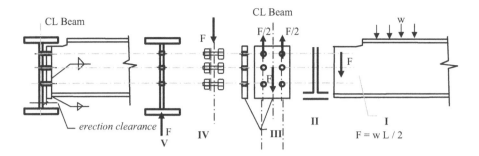

I   Supported Beam   L = Span of loaded beam
II   Fillet weld to end plate   w = Uniform distributed load
III   End / Face place   F = End shear
IV   Bolts in shear
V   Supporting Beam

Force Transfer:

FIGURE 2.3   Shear connection – with face plate.

checked for local buckling. Vertical stiffeners may be provided to prevent the local buckling of the web. The diameter and array of holes in the web should match the array of bolt holes in the outstanding leg of cleat angles.

2.1.3   BEAM TO BEAM SHEAR CONNECTION WITH FACE PLATE

Figure 2.3 shows a typical simply supported beam to beam shear connection joint with face plate welded to the supported beam (I). The reaction from the beam is transferred to the end plate by fillet weld runs (II). The end plate (III) has holes for bolts (IV), which connect the face plate to the supporting beam (V).

The difference between the cleat angle and face plate joints is that the cleats are flexible and do not allow end restraint, whereas the face plate with vertical rows of bolts may develop partial restraints. For practical purposes, moment due to such end restraint may be neglected.

The beam with a cleat angle joint is easy to maneuver at site for erection than a beam with a face plate at end. The secondary floor beams, which are generally erected after the main frame and column ties are fixed in position, should be provided with the cleated end joints.

2.1.4   BEAM TO COLUMN – SHEAR CONNECTION

Figure 2.4 describes shear connections between beam and column. In Type - a, the end reaction from the beam is transferred to the web cleats through bolts. The

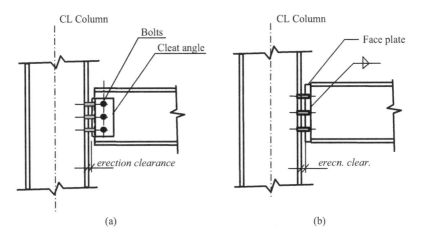

FIGURE 2.4   Beam to column – shear connection.

outstanding legs of web cleats are connected to the flange of column section by bolts that transfer the load on the column.

Type - b connection is with a face plate welded to the beam web. The face plate and the column flange are joined by bolting. The beam end reaction is directly transferred to the column through connecting bolts.

In both types, the beams are placed between the outer face of the column. So, beams can be erected with minimum erection clearance. The partial restraint in the face plate type joint may be ignored. The effective thickness of cleat angles, face plate, and the web of supported beam should be checked for bearing and block shear. The connection bolts shall be checked for single shear and bearing at column face.

2.1.5 BEAM TO COLUMN – MOMENT CONNECTION

Figure 2.5 is a popular moment connection joint with bolt. The plate girder has a long face plate. The top flange of the plate girder is butt-welded to the face plate. The rest of the contact surfaces of the girder are fillet-welded to the face plate. The depth of the girder at the end is increased by the bracket-shaped stiffener plate, which is also fillet-welded to the face plate. This is to increase the shear capacity of the girder and provide support to the deep face plate. The face plate is connected to the flange of the I-shaped column by a group of bolts in two vertical rows. A small thick plate (shear bar) is welded to the flange of the column. This is a seating stool for the girder face plate.

Actually, the depth of the face plate is guided by the numbers and array of bolts necessary to transfer the end moment and shear force to the column. The column web is strengthened by horizontal stiffeners at the location of the girder flanges joining the column to avoid local buckling. High tensile strength bolts of 8.8 grades and above should be used in this connection.

Transfer of forces from girder to column should be done by the group of bolts connecting the face plate to the column flange. The group of bolts should be designed for moment and shear. The web of the column needs to be checked for buckling at the

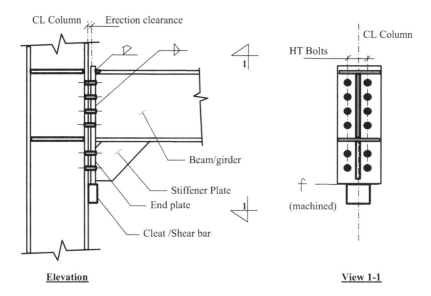

FIGURE 2.5   Beam to column – moment connection.

zone of connection. The face plate shall be checked for flexure against maximum bolt tension.

A workout example showing the step-by-step method of this moment connection is provided in Chapter 4.

2.1.6 TRUSS MEMBER JOINTS

Figure 2.6a shows elevation of a lattice girder truss. There are four panels. The top and bottom chords are parallel to each other. Diagonal members in the left two panels slope down from left to right, and diagonal members in the right panels are the same but in a mirror image. The end joints are labeled Detail A, and the joint at mid-point of the bottom chord is labeled Detail B.

Figure 2.6b shows the joint detail of a parallel chord truss connected to the flange of a column by bolts. All the truss members are made of rolled steel double angle sections placed back-to-back, separated by gusset and tack plates at intervals. The center line (center of gravity) of the top chord and the diagonal member meet at a point on the center line of the column. The ends of the top chord and the diagonal members are welded to a common gusset plate. This gusset plate is welded to a face plate, which is perpendicular to the plane of the gusset. The face plate is connected to the column flange by bolts. The bottom chord of the truss is connected to the column by bolting with a face plate and welded gusset.

Figure 2.6c shows the joint detail of a parallel chord truss, which is connected to the flange of a column by bolts. All the truss members are made of rolled steel double angle sections placed back-to-back and separated by gusset and tack plates at intervals. The center line (center of gravity) of the top chord and the diagonal member meet at a point on the outside face of the column. The ends of the top chord and

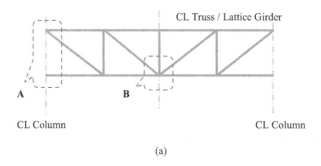

(a)

FIGURE 2.6A Elevation.

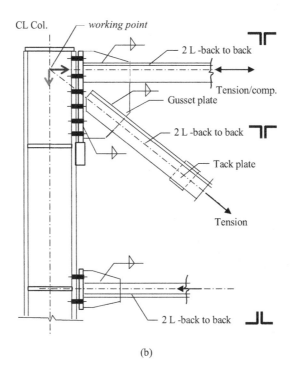

(b)

FIGURE 2.6B Detail A – Type 1.

the diagonal member are welded to a common gusset plate. This gusset plate is welded to a face plate, which is perpendicular to the plane of the gusset. The face plate is connected to the column flange by bolts. The bottom chord of the truss is developed with the welded gusset and a vertical end plate, which is joined to the column flange by bolts.

Figure 2.6d shows a view of the face plate used in Figures 2.6b and c. The face plate has two vertical rows of holes for installing bolts to connect with the column flange. The face plate is resting on a seating bar welded to the face of the column.

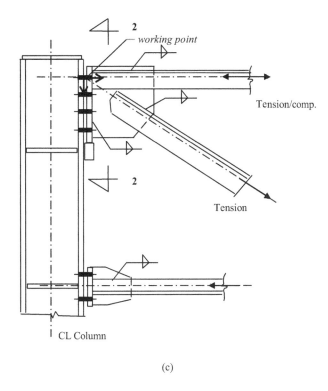

(c)

FIGURE 2.6C   Detail A – Type 2.

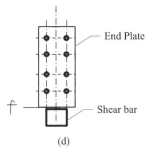

(d)

FIGURE 2.6D   View 2–2.

Figure 2.6e is an end connection detail of the roof truss. Here, the top chord of the truss and the diagonal member meet at an elevation just above the column cap level. This joint is a welded joint developed by a common gusset plate resting on a sole plate meeting at a right angle to each other. The sole plate is placed on the column cap plate and held in position by bolts. The end reaction of the truss is transferred to the top of the column. The bottom chord of the truss fabricated with a gusseted face plate is joined to the flange of column by a bolt.

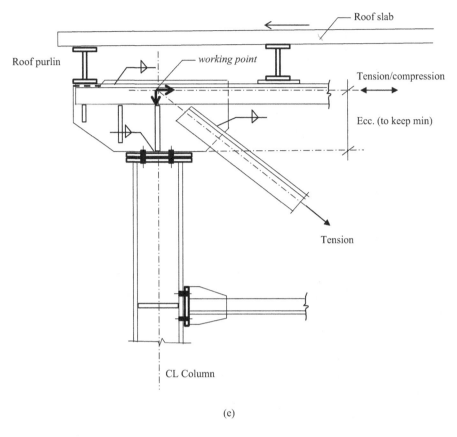

(e)

FIGURE 2.6E   Detail A – Type 3.

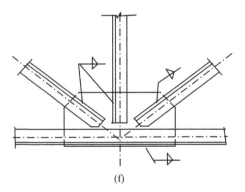

(f)

FIGURE 2.6F   Detail B.

Figure 2.6f is a typical welded joint of all members meeting at a point on the center of the gravity line of the bottom chord of the truss. The diagonals and vertical chord member is welded to the bottom chord by a common gusset.

All the above sketches show the common types of truss end connections used in industrial structure. The concept is to fabricate the truss (by welding) at shop and join to the supporting column with bolts (HT bolts). If the span of a truss is more than allowable transportable length (~13 meter), the truss is fabricated in pieces and assembled at the job site before erection. These field joints may be bolted (by HT bolt) or site-welded on the ground.

The joint shown in Figures 2.6b and c are similar, but the load center of gravity (C.G) lines meet at a different location. In Detail A – Type 1, the CG lines of truss chords intersect at column center line, so there would not be any additional moment in member design due to eccentricity at the joint. However, the vertical length of the gusset is increased to accommodate the position of the diagonal member.

There would be an additional moment at the top of column in Detail A – Type 2 due to the eccentricity of the working point, where the load CG of the diagonal member and the top chord of the truss meet at a point outside the column face.

The end connection shown in Detail A – Type 3 is suitable for a large span roof bearing heavy dead load from cast-in-place roof slab. In this type of connection, the intersection point of load CG should be brought close to the column cap level by adjusting the slope of end diagonal. The bolt group at column cap level should be designed for the moment due to eccentricity of the horizontal component of force above the column cap.

The designer should note that the weld length of diagonal member, diameter of connection bolts, and sizing of face plate are the critical items in the design of truss end connection. The designer must carefully select the magnitude and direction of member loads for all critical load combination cases and calculate the resultant forces transferred to the column at this joint. The group of bolts joining the face plate to the column should be checked for moment and shear. The face plate would be designed for bending against maximum bolt tension.

2.1.7 COLUMN BRACING JOINTS

Figure 2.7a shows a typical column bracing joint at floor beam level. The floor beam connects to the column web by a pair of angles cleat-bolted to the column flange. There are two diagonal bracing of double angle section meeting at the corners of the beam-column joint at the top and bottom level of the floor beam. The load CG of bracing members and the flanges of the beam coincide at the face of the column. The end connection between bracing, column, and flanges of the beam at the top and bottom is achieved by field bolting. A fabricated piece of component made with a vertical gusset plate, two face plates welded at right angles for corner joint, and bearing holes for bolts are the jointing medium for bracings. The bolts shall be checked for shear force and moment, as applicable.

Figure 2.7b is a joint between two Vertical bracings on the top and bottom of a horizontal tie member meeting at the face of the column web. A common vertical gusset plate is welded to the column web at a right angle to the surface of the web to receive all three members meeting at the joint. This gusset plate has an array of matching bolt holes for connection with the bracing members on the slope and the horizontal tie. All members are joined with the gusset plate by permanent bolts. The bolts should be designed for shear against member load.

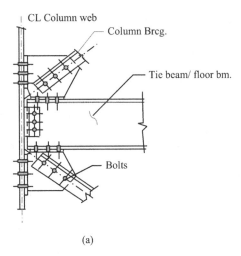

(a)

FIGURE 2.7A   Column bracing.

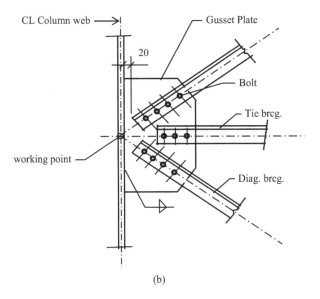

(b)

FIGURE 2.7B   Column bracing.

Figure 2.7c is a joint between two diagonal bracings on the top and bottom corner of a floor beam connected to a column web plate. There is a RCC floor slab supported on the floor beam. The top bracing is joined by bolting with a vertical gusset plate welded perpendicular to the face of the column web plate. This gusset is shaped matching the slope of the bracing member; it is not connected to the slab and floor beam. The end of the floor beam is fabricated with a deep face plate and a knee gusset welded to the face plate and bottom flange of the floor beam. Bolt holes are provided

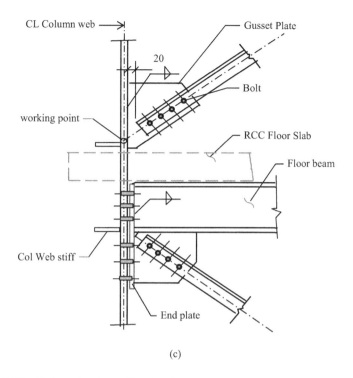

CL Column web →|

Gusset Plate

20

Bolt

working point

RCC Floor Slab

Floor beam

Col Web stiff

End plate

(c)

FIGURE 2.7C Column bracing at floor level.

on the face plate and in the knee plate to join with the column and connect the diagonal bracing at bottom. All field connections are bolted.

The design of the weld connecting the gusset to the column above the floor should
be checked for eccentric moment and shear. The bolts joining the member to the gusset should be checked for shear against member force. The joint below the floor
needs to be designed for beam end forces and the bracing member force. The forces
are to be resolved and the bolts group at the column faces must be checked against
resultant load. The design of the face plate should be for flexure and shear.

Figure 2.7d is a joint detail of diagonal bracings meeting at the bottom flange of a
beam. A wide vertical gusset plate welded to the bottom flange is provided to connect
the diagonal bracing members. There are holes in the gusset matching with slopes of
the bracing member. The bracing members are joined with the gusset plate by bolting.

Column bracings are used to resist lateral forces on the building and provide stability of framework against sway. The bracing members connect corners of all beam
column joints in the bracing bay at the vertical plane. Bracings are also provided in
the horizontal plane. Horizontal bracings connect the roof truss joints at the rafter
and tie level, top flanges of floor beams in plan, walkway, and platform members, etc.
The cutting length and slope of bracing members as shown in workshop drawings are
often found mismatching with actual measurements available at site. So, the bracing
joints are one of the critical items that should be developed, giving some flexibility
and provisions for adjustment during erection at site.

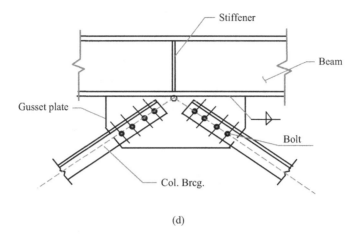

(d)

FIGURE 2.7D Column bracing with beam.

In general, the field connection should be bolted joints for bracings. Field-welded joints are also used for bracing connection.

2.1.8 SPLICE JOINT

Figure 2.8a shows the splice joint of a plate girder. This is a cover plate splice with bolts holding the flanges and web of a plate girder separately. There is a gap of 3 mm between girder ends at the splice location. These cover plates span over the gap and maintain continuity of member properties over the joint. The joint detail is symmetrical about the center of splice. These cover plates are on both sides of the flanges and web. The bolt holes in cover plates and the girder section are drilled together on assembled position to avoid misalignment of bolts.

Figure 2.8b shows the splice joint of a rolled steel I beam. This splicing is made by cover plates and bolts holding the flanges and web of the rolled steel beam. There is a gap of 3 mm at the splice location. These cover plates span over the gap and maintain continuity of member properties over the joint. The joint detail is symmetrical about the center of the splice. Two webs cover plates with an equal number of bolts on either side join the webs of separated sections. The flange cover plate is a long plate with bolt holes, with the width same as the width of a rolled steel member or more. A tapered washer is used underside of rolled steel flange matching its inclined face and to give perfect bearing of the nut. The bolt holes in cover plates and beam are drilled together on assembled position to avoid misalignment of bolts.

Figure 2.8c shows the splice of a two rolled steel angle section placed back-to-back with a separator plate between connecting legs. There are two bend plates at the inner face of angles and one wide plate over outstanding legs, used as a cover plate in splicing. The jointing is done by bolts.

Figure 2.8d shows the splice joint of a column. This splicing is made by cover plates and bolts holding the separated flanges and web of the member section. There is a gap of 3 mm between girder sections at the splice location. These cover plates span over the

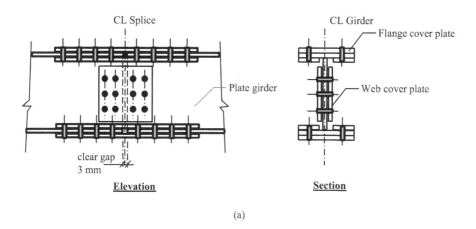

(a)

FIGURE 2.8A   Splice in plate girder.

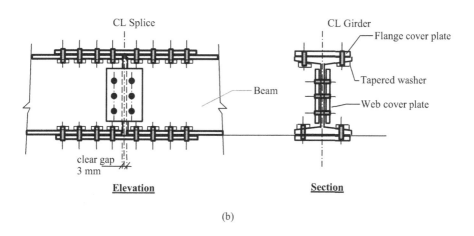

(b)

FIGURE 2.8B   Splice in rolled steel beam.

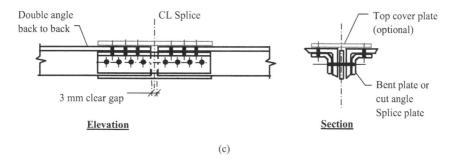

(c)

FIGURE 2.8C   Splice in rolled steel angle.

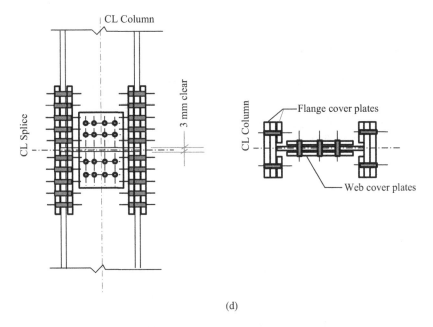

(d)

FIGURE 2.8D   Splice in column.

gap and maintain continuity of member properties over the joint. The joint details are symmetrical about the center of splices. The bolt holes in cover plates and the girder section are drilled together on assembled position to avoid misalignment of bolts.

Splices should preferably not be located at points of maximum stress. The area of the splice plates should not be less than 5% in excess of the area of the element spliced; their center of gravity should coincide, as closely as possible, with that of the element spliced. There should be enough bolts on each side of the splice to develop the load in the element spliced plus 5%, but in no case should the strength developed be less than 50% of the effective strength of the materials spliced. Location of the column splice should be selected at 1.2 ~ 1.5 meters above floor level. This is suggested to get easy access for bolt tightening standing on floor beams. More details of column splices are shown in Chapter 6.

Design of the splice should be done for full strength of the member section unless noted otherwise. Each component of a splice joint should be designed independently. For example, in a plate girder splice, the cover plates and connection bolts at the flange joint must be capable of transferring the flange force from one part to the other across the splice. Similarly, the web cover plates and connecting bolts should be designed for web shear. If there is a group of bolts, it should be checked for shear and moment due to eccentricity between splice center and CG of the bolt group. Workout examples are given in Chapter 4.

2.1.9 PURLIN AND SIDE GIRTS

Figure 2.9 shows a roof purlin (joist section) seated on the top chord of a truss. The web of the purlin (joist) is held in position by a bracket-shaped bent plate welded to the top of the truss member and bolted to the purlin. There is a detail of a cut joist bracket alternate to bent plate.

Figure 2.10 shows two side-girt joint details – one with channel and the other is a joist section being used as a side girt. Girt members are resting on cleats welded to the column face. Bolts used in girt member and seating cleat joint are of 4.6 grade ordinary mild steel bolts.

The roof purlin and side girts shown in the above figure are a rolled steel section. The bolts for joining such secondary members should be 4.6 grade ordinary mild steel. Side girt (runner) and purlin members made of cold-formed thin metal section (Z or C section) may be connected with similar seating cleats but with thickness limited to 5 mm. This is to allow full penetration of fixing screws used in thin metal connection.

2.1.10 COLUMN BASES

Figure 2.11 shows a built-up I-section column resting on a rectangular base plate. The column shaft is attached to the base plate by a gusset plate and stiffener by a fillet-weld joint. The base plate sits over a grout cushion laid on the RCC pedestal.

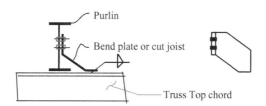

FIGURE 2.9   Purlin connection.

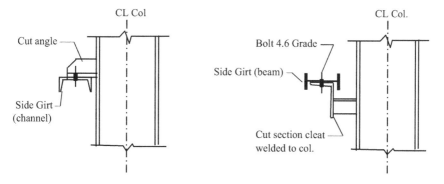

FIGURE 2.10   Side girt connection.

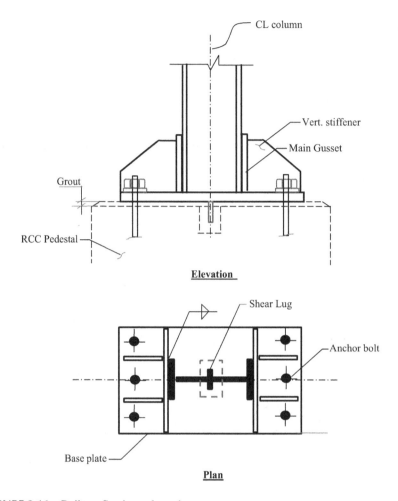

FIGURE 2.11   Built-up Section column base.

The column base is fixed in position by an anchor bolt and shear key embedded into the concrete pedestal.

Figure 2.12 shows a wing plate type column base. The main column shaft sits on a smaller size base plate resting on grout cushion over a reinforced column pedestal. A wide and thick top plate is fixed at a higher level (same as column width) parallel to the base plate. This top plate is butt-welded to the column and resting over vertical stiffeners spread like wings above the base plate. The anchor bolts are raised above this thick top plate with threaded length projecting above top plates. Double nuts are provided to fasten the bolt and tighten against the top plate. All joints are welded.

Figure 2.13 is a plan view of a common base plate holding a twin legged (I-section) column. The base plate, main gusset plates, stiffener plates, and anchor bolts are symmetrically placed. All the gussets and stiffeners are fillet-welded to the base plate and column section.

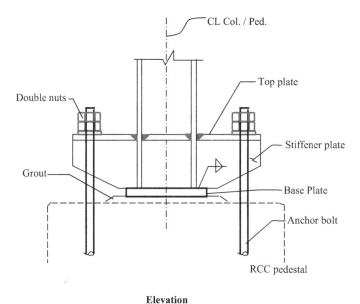

Elevation

FIGURE 2.12   Wing plate type column base.

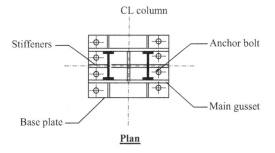

Plan

FIGURE 2.13   Twin legged column base.

The details of column bases shown above are commonly used in industrial plant building and pipe racks. Figure 2.11 is a general sketch for a single shaft column. These types of column bases are suitable for pipe rack structure. For crane bearing buildings, I-shaped columns and bases of larger size are very popular compared to twin-based columns shown in Figure 2.13. The twin-shaped column with lacing bars is good for controlling lateral deflection, but it involves more time and cost in fabrication. Figure 2.12 shows a wing type base plate, which is good for bases with large moments (seismic or wind) and compared to axial forces.

The size of the base plate, numbers, and size of anchor bolts and stiffeners vary according to the loading at the column base. The reader should note that the design capacity of anchor bolts should be checked with the concrete break-out strength or rupture of the pedestal over the embedded length (refer to ACI 318 R-14 Building Code requirements for Structural concrete and commentary).

Workout examples for column base design are provided in Chapter 3.

2.1.11 JOINT WITH CONCRETE STRUCTURE

Figure 2.14 shows a connection between a steel beam and a reinforced cement concrete (RCC) beam. There is a mild steel plate with welded lugs embedded into the concrete beam as shown, while casting. The surface of the embedded steel plate is exposed to air, matching with the vertical side of the supporting RCC beam. The supported steel beam is joined to the steel plate by a cleat angle, which is welded to the steel plate and bolted to the web of the steel beam.

The above sketch is a typical arrangement of a steel member joint to a reinforced cement concrete (RCC) section. Here, the load is transferred from the steel beam to the connecting bolt and cleat. The outstanding flange of cleat is welded to the mild steel insert plate at site. The holes in the web of the beam may be slotted to provide erection tolerance and allow movement for differential settlement between members, if any. This is a simply supported connection, so lugs attached to the insert plate and its connecting weld are to be designed for shear. Similar types of joints are used for connecting members with RCC columns and other structure. The mild steel plate and the group of lugs are designed for moments and shear acting at the joint.

2.2 WELDED JOINTS

Many types of welded joint details are available in textbooks and design manuals. In this chapter, the reader has been introduced to selected types of joints that are currently used in projects at different countries and easy to construct.

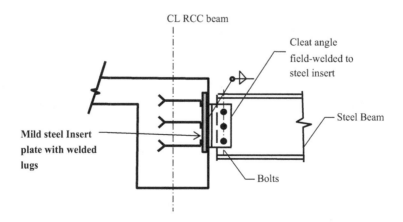

FIGURE 2.14   Steel beam to concrete beam joint.

2.2.1 Fillet Weld Joint

Figure 2.15 is a typical fillet-welded joint used to connect a vertical column bracing member with a column section at the erection site. The column section is provided with a fabricated gusset plate, welded to the outside face of the column and projecting at a right angle to the flange plate. The shape of the gusset is tapered towards the free end, matching the width of the connecting bracing member.

The bracing member is a double angle section placed back-to-back and separated by tack plates equal to the thickness of the column gusset. The gaps between the legs are kept free at both ends so that the gusset plate can fit into the slot during erection. The connecting legs of bracing and the gusset plate need to have drilled holes for erection bolts (14-mm bolts are generally used) at same elevation and spacing according to the workshop drawing. After erecting the bracing member in position and aligning the same to correct level, the erection bolts are tightened to bring the contact surfaces between the legs and the gusset plate close together. There should not be any gap beyond the permissible limit for fillet welding (should not be more than 1/16 inch), as per code. Finally, this site welding should be done on erected position. This joint is a combination of shop and field weld. The design examples are available in Chapter 3.

2.2.2 Butt Weld Joint

Figure 2.16 shows basic types of butt-weld joints that cover the requirements of general building and pipe rack structure.

Details of square-butt welding are on top left. Here, the plates meet at a plane, leaving a gap of about 3 to 5 mm, which is called root gap. A small metal strip has been provided on the back side covering the root gap. This back strip holds the metal between joining faces during welding. Welding is done from the top, filling the square gap and ending with an oval-shaped deposit at top. The square butt weld is generally used to join thin metal sheets, e.g., welding of liner plates, chequered plate flooring, etc.

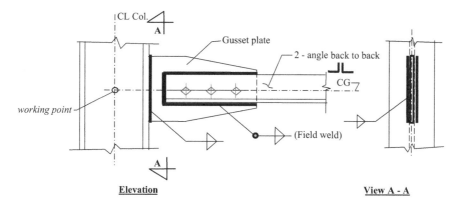

FIGURE 2.15 Fillet weld joint.

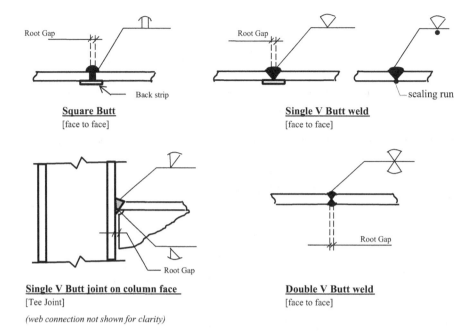

FIGURE 2.16 Butt weld joint.

The sketch on top right is a Single V Butt weld joint between two plates meeting at the same plane. The edges of both the plates are bevel-shaped, forming a V-shaped grove at the joint. A thin metal strip has been provided underside of the joint to prevent flow of material backside.

In the process, the welding metal is deposited inside the grove by multiple passes with an oval-shaped formation at top. This is how the single V joints are welded. The back strips are not provided in all cases. The welder fills up the root with the first run and completes the rest of the welding. After that, the root is cleared by back gouging (the process of removing weld and base metals from the backside of a weld for cleaning), and a sealing run is provided. This cleaning eliminates impurities in the root weld, done without back strip. In a normal process, the impurities come out of the molten pool of weld metal towards atmosphere as a slug. So, the root run without a back strip needs cleaning and subsequent sealing.

The sketch on bottom right is a double V butt joint. In this joint, the edges of the two plates form a V-shaped groove–near side and far side. The edges of the plates are beveled from both sides. There a root gap of about 3 mm between edges at center. The welding is done from both sides to form a metal deposit of double V shape.

The sketch on left in the bottom row is a Tee-butt joint between the column face and the top flange of a plate girder. The top flange of the girder is beveled to form a single V, whose vertical side is the column face. The welding is done from the top, filling the V-shaped groove.

The joint designer should note that a butt weld joint is considered as a full strength joint, because the strength of weld metal is higher than the strength of parent metals being joined. But it depends on the correct edge preparation, appropriate size and type of electrode, compliance of preheating requirement, welding accessibility, environment (shop or site), and quality of welder (for manual welding). It may not be possible to provide an ideal condition for welding at site, hence a strength reduction factor (~80%) is applied for fillet and butt welding done at site. The designer should select the appropriate type of edge preparation to ensure that joints do not suffer from lack of penetration or lack of fusion. Experienced welders suggest single V edge preparation for plates up to 16 mm thick and double V for plates of higher thickness. However, it also depends on the accessibility for welding.

The root gap between components is an important factor and to be provided as per guidelines given in welding codes and AISC manual. There are cases of joint failure due to excessive gaps and defective welds. The joint designers should be aware of facility available for welding accessibility and accordingly select the proper method and type of joint.

2.2.3   BEAM TO BEAM CONNECTION

Figure 2.17a and b show simply supported shear connections welded at site. These connections are commonly used for floor beams and pipe rack structure. There are two types of connection details. Figure 2.17a shows the connection between two beam sections by cleat angles. The supported beam has wired-on cleat angles held by erection bolts. The cleat angles have connecting holes at outstanding legs, matching with the holes provided in the web of the supporting beam. The pair of cleat angles joins the beams in position by erection bolts and then field-welded.

Figure 2.17b is a joint of two beams by face plate. The face plate is shop-welded to the end of a supported beam at a plane perpendicular to the web. The end plate and

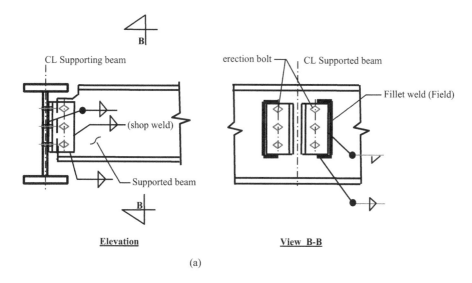

(a)

FIGURE 2.17A   Shear connection with cleat – welded.

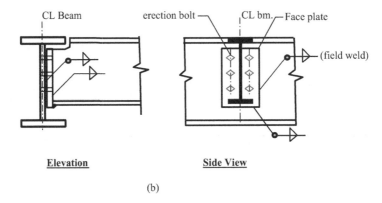

(b)

FIGURE 2.17B   Shear connection with face plate – welded.

the supporting beam web have matching bolt holes. The face plate is held in position by erection bolts and then welded all-round at site.

The connection with wired-on cleat angle has more flexibility than face plate type because it provides more erection clearance than the latter. The beams are held in position by the erection cleats until the leveling and alignment of framework are complete.

2.2.4   Beam to Column – Shear Connection

Figure 2.18 shows shear connections between beam and column. The sketch on left joins the beam with column by a pair of web cleats. These cleat angles are wired-on the beam web by erection bolts. The outstanding legs of these cleat angles and

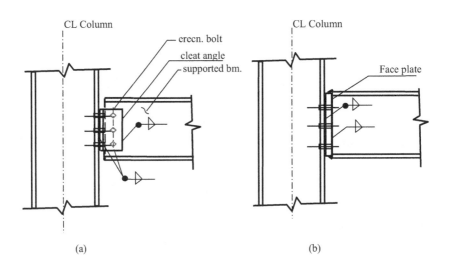

(a)   (b)

FIGURE 2.18   Beam to column shear connection by welding.

supporting column flange have matching holes for erection bolts. After erection, the cleat angles are welded to the web of beam and the column flange.

The figure on the right-hand side is a beam joined to the column flange through an intermediate face plate. The face plate is welded to the beam end at a right angle to the web plate at shop. There are holes in the face plate and the supporting column flange for holding the components by erection bolts. The face plate is finally site-welded to the column flange. These connections are suitable for transferring vertical shear and axial compression of a simply supported beam to column framework.

2.2.5   BEAM TO COLUMN – MOMENT CONNECTION

Items:
1. Plate girder/rolled steel joist.
2. Top cover plate or moment plate.
3. Bottom cover plate or moment plate.
4. Web cleat (wired on erection bolt in slotted hole).
5. Bracket or vertical stiffener.
6. Column stiffener – horizontal rib plate.
7. Column stiffener – diagonal rib plate.
8. Column.
9. Back strip.

Weld Details:
S1: Fillet weld at site – top cover plate to top flange of plate girder.
S2: Fillet weld at site – bottom flange of plate girder to bottom cover plate.
S3: Fillet weld at site – cleat angle to web plate of girder and column.
S4: Fillet weld – stiffener plate to column face and bottom cover plate.
S5: Single V butt weld at site with backing strip – top cover plate to column flange.
S6: Double V butt weld – bottom cover plate to column flange.

The Figure 2.19 is a detail of a welded moment connection joint between a plate girder and column. The plate girder transfers moment to the column by two moment (cover) plates welded at site. The bottom cover plate is attached to the column flange by butt weld at shop. It is supported by a vertical web plate forming a bracket on the face of column. There are two web cleats in the plate girder for transferring vertical shear. One cleat is shop-welded to the column, face and the other would be site-welded (Figure 2.19).

The holes for the erection bolts in the web are slotted to allow lateral shift during alignment. The top cover plate is supplied as wired-on item. After erection, leveling, and alignment of the plate girder on the column bracket – the top cover plate is butt-welded to the column flange and fillet-welded on the top flange of plate girder – the bottom flange of the plate girder is fillet-welded to bracket and the cleats are also fillet-welded to the column face and plate girder web. The column web is strengthened by horizontal and diagonal ribs.

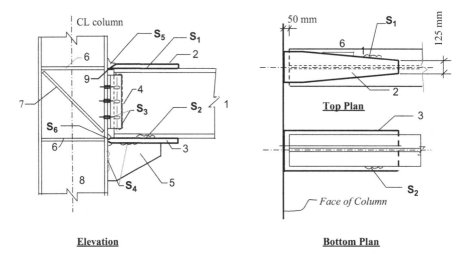

FIGURE 2.19   Beam – column moment connection joint – welded.

This type of moment connection has adequate flexibility for site weld construction. Other than the top flange, all welding works are fillet welds, which are easy for site construction.

2.2.6   BRACING JOINTS

Figures 2.20 to 2.22 are standard column bracing connections used in plant buildings at pipe rack structure.

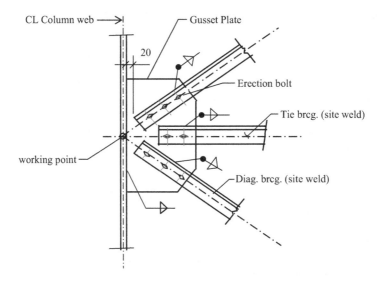

FIGURE 2.20   Diagonal bracing – welded.

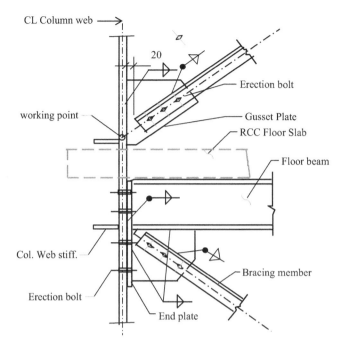

FIGURE 2.21   Column bracing joint at floor level – welded

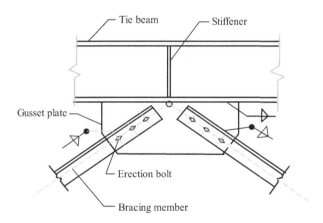

FIGURE 2.22   Bracing joint at center of tie beam – welded.

Figure 2.20 shows a vertical bracing joint having two diagonal members, one at top and the other at the bottom of a horizontal member meeting at a point on the face of a column web. The ends are connected to a common gusset, which is shop-welded to the column web. The bracing members are fixed with erection bolts and then welded at site.

Figure 2.21 shows a joint of two diagonal bracings and a floor beam meeting with a column web on a vertical plane. The diagonal bracing at top is connected with a single gusset welded to the column web above the floor slab. The bracing member is fixed with erection bolts and then field-welded. The floor beam is fabricated with a deep face plate projecting below its bottom flange. A vertical gusset serving as bracket plate is joined to the face plate and soffit of the beam. The beam face plate is fixed to the column web by erection bolts and then site-welded. The diagonal bracing at bottom is joined with the bracket web by erection bolts and finally site-welded.

Figure 2.22 shows two diagonal bracings joining to the mid-point of a tie beam in a vertical plane parallel to the web of the tie beam. A common gusset plate is welded to the soffit of the beam. The bracing members are diagonally opposite (mirror imaged). These members are fixed with the gusset plate by erection bolts and then welded at site.

2.2.7 SPLICE JOINTS

The sketches shown below are the common details used for welded splice at construction site and workshop. The splice used in workshop is generally of full penetration butt-welded joints because facilities like preheating and other related work are easily available in workshop. The site joints should be fillet-welded type as far as possible with cover plates.

Figure 2.23 shows a splice joint of a plate girder. The members to be joined are aligned face to face leaving a gap of 3 mm at center line of joint. Cover plates are provided on both sides of the flanges and the web, which span over the gap and then welded to column sections at all sides around the contact surfaces. Erection bolts are tightened to bring the spliced components in close contact during welding. All the cover places are symmetrically placed at the center of splice.

Figure 2.24 is the splice of a rolled steel member jointed by butt-welding the flanges. The butt weld is double-V, welded from the outside face of the flange. The webs are joined by cover plates and fillet welding. There is a coping hole drilled in the web at the junction point of beam web and flange. This is to give space for

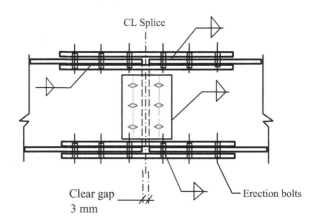

FIGURE 2.23 Plate girder splice – welded.

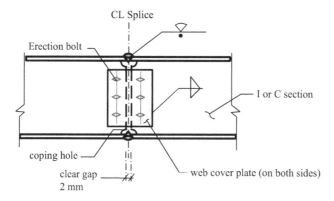

FIGURE 2.24   Rolled steel section – welded splice.

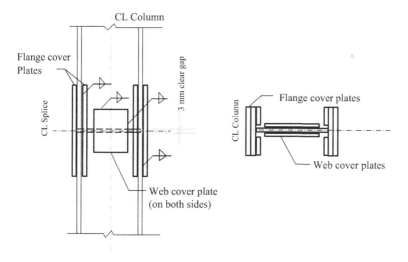

FIGURE 2.25   Column splice – fillet weld. (Erection bolts are not shown for clarity.)

continuing sealing run over the root. The web cover plates span over the gap between beam ends and are fillet-welded.

Figure 2.25 shows a column splice detail with cover plates and fillet weld. The components are aligned face to face leaving a gap of 3 mm. The column sections are held in position by cover plates and erection bolts. The cover plates provided on both sides of the flanges and web span over the gap and are then fillet-welded to the column section around contact surfaces. All the cover plates are symmetrically placed about the center of the splice and tightened by erection bolts to bring joining surfaces in close contact.

Figure 2.26 is a column splice joint, where the contact surfaces of the column shafts are milled or machined to ensure perfect contact for bearing. In addition, extra cover plates are provided outside the flanges and both sides of the web to transfer

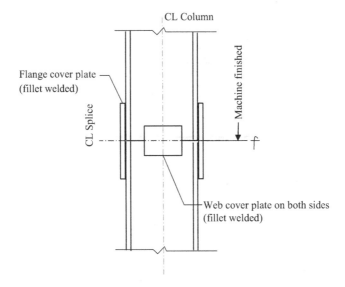

FIGURE 2.26 Column splice – contact faces machined for bearing. (Erection bolts are not shown for clarity.)

moment and share a part of axial force. For example, if the milled surface ensures 60% of contact area for bearing axial load, the rest will be shared by cover plates.

However, welded splices should be done on ground or at shop. All field splices aloft should be bolted as far as possible because welding at higher altitude would not be possible. Friction grip bolted connections are preferred to make it a non-slip type joint, which is equivalent to a welded continuous member.

2.3 DEFECTIVE DESIGN OF JOINTS

There are many incidents of structural steel failure. Reasons and evidence for such failures are generally due to overloading of structure, defects in material, design defects, wrong construction, and failure of joints, sometimes beyond conclusion. Following are some examples of joint failure due to design defects in detail engineering work.

2.3.1 ROOF TRUSS JOINTS

Figure 2.27 shows a detail of roof truss joints followed in construction of an industrial plant building. The entire roof structure collapsed within a few weeks after the construction of the roof slab. Actually, the truss chords and diagonal members sheared off from the surface of the column flange, including the connecting plate and cleat angle.

From the sketch below, the reader can see that the roof leg of the column is a rolled steel beam section. The roof truss has two parallel chords. The top chord is supporting roof purlins bearing a reinforced concrete roof slab. This top chord is a

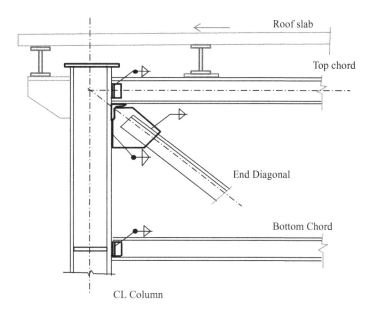

FIGURE 2.27   Wrong detail – Roof Truss connection.

rolled steel beam joined by a pair of welded cleats to the column flange. There is a seating cleat welded to the column face used for resting the top chord beam. The diagonal chord is a double angle section, which is connected to the face of the column flange by a single vertical gusset plate fillet-welded at site. The bottom chord is a rolled steel beam joined to the column face by a pair of web cleats. Connection between truss members and column faces are all site-welded joints.

The evidence showed that the gusset plate connecting the diagonal member to the column face by two vertical runs of fillet weld was torn off from the face of the column. So, the entire roof structure collapsed.

(The reader should follow the details of roof truss connections given in Section 2.1.6; all field joints aloft should be bolted connections.)

2.3.2   Hopper Plate Joint

The sketches shown below are substitute details of hopper plate joints:

a) As in drawing
b) As constructed

This was a case of a raw coal bunker that collapsed a few years after being put into service.

Figure 2.28(a) is the detail of a hopper plate joint as per drawing. The inclined plates are held in position with the help of a wide backing strip or plate placed behind the hopper plate. The center of the backing plate is the center of joint. This back plate is welded to the top hopper plate and half of its width is projected below the line of

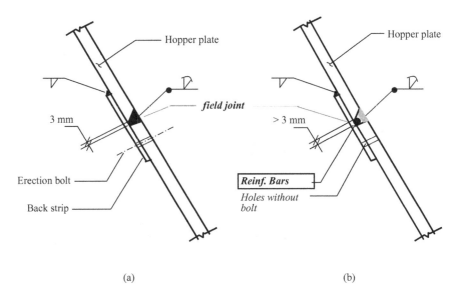

FIGURE 2.28 Wrong construction – Hopper plate.

joint. There are holes in this projected part for erection bolts at intervals. The bottom hopper plate is hanging from this back strip, keeping root gap along center of joint and held in position by the erection bolts. The root gap is 3 mm along the line of joint. After erection at site, the bottom plates should be aligned by tightening erection bolts and then welded to the top plate, as shown in drawing. This is as planned in drawing detail.

Figure 2.28(b) on the right showed what actually happened in construction, which was seen in the collapsed structure on the ground. There was no erection bolts, and the gaps between hopper plates were more than that shown in the drawing. A piece of reinforcement bars was placed at root to fill up the excess gap, and the rest of the grove was filled with weld deposit. At some places, the bottom plate was welded to the back strip only.

REFERENCES

1. Steel Construction Manual AISC – Fourteenth edition.
2. IS 800: 2007 General Construction in Steel – Code of Practice.

3 Welded Joints
Workout Examples

3.1 BEAM TO BEAM – SHEAR CONNECTION

In this section, workout examples of welded shear connections with web cleats and face plate are provided, in accordance with AISC – ASD and LRFD; Indian Standard (IS) code- working stress and limit state.

3.1.1 WEB FRAMING CLEAT ANGLE – IS 800 – WORKING STRESS

a) Sketch

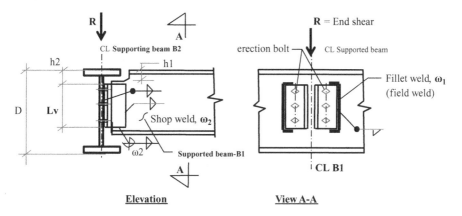

FIGURE 3.1 Shear connection with angle cleats.

b) Materials

TABLE 3.1
Properties of Structural Steel Materials

Grade			E 250 (Fe 410 W) A				Reference
Yield Stress	fy	250	240	230	MPa	IS 800:2007 – Table 1
d or t		<20	20–40	>40		
Tensile Stress	fu	410	410	410	MPa	
Elasticity	E		200000		MPa	IS 800:2007 – 2.2.4.1
Material Safety factor	γ_{m0}	1.10	γ_{m1}	1.25		IS 800:2007 – Table 5
Electrode Ex40-44xx			fuw	540	MPa	IS 814:2004 – Table 6
	γ_{mw}	1.25	fyw	330	MPa	

DOI: 10.1201/9781003539124-3

c) Member Description

Supported member:	B1	ISMB 600	D = 600 mm
Supporting member:	B2	ISMB 600	D = 600 mm

d) Design Load

End Shear, R = 70% of web shear capacity. (Assumed for Joint design)
(Maximum reaction that can be produced by the sectional modulus of the
beam for uniform distributed load is generally less than 50% of its web shear
capacity. Hence, it is advised to design the connection for reactions at least
over 50% of its web shear capacity, unless the actual design force is provided.)
Web shear, beam B1:

$$\text{Effective depth of web}, D - h1 = 555\,\text{mm}; h1 = 45\,\text{mm}$$

$$\text{Web thickness}, t = 12\,\text{mm}; h2 = h1 + 15 = 60\,\text{mm}$$

$$\text{Shear stress}, \tau sh = 0.4 fy = 0.4 \times 250 = 100\,\text{Mpa}. \left(\text{IS}\,800:2007 - 11.4.2\right)$$

$$R = \text{End shear} = 0.7 \times \left(D - h1\right) \times t \times 0.4 fy / 1000$$
$$= 0.7 \times 555 \times 12 \times 0.4 \times 250 / 1000 = 466\,\text{kN}.$$

Cleat sizing: Provide web cleat angles – 2 ISA 100 100 12

$$\text{Length of cleat}, Lv = 480\,\text{mm}$$

$$\text{Leg thickness}, Lt = 12\,\text{mm}$$

$$\text{Leg size}, Lh = 100\,\text{mm}$$

$$\text{Shear capacity of cleats} = 2 \times 480 \times 12 \times 100 / 1000 = 1152\,\text{kN}. > R : \text{Okay}.$$

e) Fillet Weld Sizing

(Refer "Design of Welded Structures – O W Blodgett".)
Field Weld to Supporting Beam, ω_1:
It is assumed that the two angles bear against each other for a vertical distance equal to 1/6 th of their length. The remaining 5/6 th of the length is resisted by the connecting welds. It is also assumed that these forces on the welds
increase linearly, reaching a maximum (fh) at the bottom of the connection.

$$P = 1/2\left(\text{fh}\right)\left(5/6\,Lv\right)$$

$$\text{fh} = \left(9RLh\right) / \left(5Lv^2\right)$$

$$\text{Vertical force on weld}, fv = R / 2Lv$$

$$\text{Resultant force on the weld}, fr = \sqrt{\left(\text{fh}^2 + \text{fv}^2\right)}$$

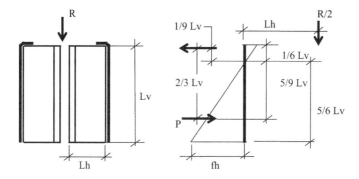

FIGURE 3.2 Welding of cleat on supporting beam web.

$$R = 466 \text{ kN}; \text{fh} = \left(9 \times 466 \times 100\right) / \left(5 \times 480^2\right) = 0.364 \text{ kN / mm.}$$

$$\text{Lh} = 100 \text{ mm}; \text{fv} = 466 / \left(2 \times 480\right) = 0.485 \text{ kN / mm.}$$

$$\text{Lv} = 480 \text{ mm}; \text{fr} = \sqrt{\left(0.364^2 + 0.485^2\right)} = 0.606 \text{ kN / mm.}$$

Provide fillet weld of size, $\omega_1 = 10$ mm.

Permissible stress in fillet weld, fw $= 125 \text{ MPa}; \left(0.4 \text{ fy}\right) \left(\text{IS } 800 : 2007 - 11.6.3\right)$

EX40XX–EX44XX electrode
Strength of fillet weld connecting cleat angles to the web of supporting beam at field

$$= 0.8 \times 0.7 . \omega_1 . \frac{\text{fw}}{1000} \text{ kN / mm} \left(20 \text{ \% stress reduced for site weld}\right)$$

$$= 0.8 \times 0.7 \times 10 \times \frac{125}{1000}$$

$$= 0.700 \text{ kN / mm} > \text{fr}; \text{Safe.}$$

Shop weld of cleat to supported beam, ω_2:

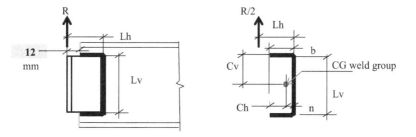

FIGURE 3.3 Shop welding of cleat on supported beam web.

In Figure 3.3,

$$n = b^2 / (2b + Lv)$$

$$Ch = Lv - n - 12$$

$$Cv = Lv / 2$$

$$b = Lh - 12$$

$$Jw = \left[(2b + Lv)^3 / 12 - b^2 (b + Lv)^2 / (2b + Lv) \right]$$

Twisting (horizontal)

$$fh = TCv / Jw = \left[R(Lh - n) Cv / 2 Jw \right]$$

Twisting (vertical)

$$fv_1 = TCh / Jw = \left[R(Lh - n) Ch / 2.Jw \right]$$

Shear (vertical)

$$fv_2 = \left[0.5 R / (2b + Lv) \right]$$

$$fr = \sqrt{ \left[fh^2 + (fv_1 + fv_2)^2 \right] }$$

In this example,

$$b = 88 \, mm \, (100 - 12); R = 466000 \, kN$$

$$Lv = 480 \, mm$$

$$Lh = 100 \, mm$$

$$n = 11.8 \, mm \, (88^2 / (2 \times 88 + 480))$$

$$Ch = 76.2 \, mm \, (100 - 11.8 - 12)$$

$$Cv = 240 \, mm \, (480 / 2)$$

$$Jw = 23525035 - 3808537 = 19716498 \, mm^3$$

$$fh = 250 \, kN / mm$$

$$fv1 = 79\,kN/mm$$

$$fv2 = 355\,kN/mm$$

$$fr = \sqrt{\left[250^2 + \left(79 + 355\right)^2\right]} = \mathbf{501}\,kN/mm.$$

Provide fillet weld size, $\omega_2 = 6\,mm$

Strength of fillet weld connecting cleat angles to the web of supported beam at shop:

$$= 0.7\omega_2 fw\;N/mm = 0.7\times6\times125\;N/mm = 525\,kN/mm > fr; Safe.$$

3.1.2   WEB FRAMING CLEAT ANGLE – IS 800 (LIMIT STATE)

i)   **Sketch** – See Figure 3.1
ii) **Materials** – Table 3.1
iii) **Member Description**

Supported member:	B1	ISMB 600	$D = 600\,mm$
Supporting member:	B2	ISMB 600	$D = 600\,mm$

iv) **Design Load**

$$\text{End shear}, R = 0.7\times\left(D - h1\right)\times t\times\left(fy/\sqrt{3}\right)/\gamma_{m0}\;\left(IS\,800:2007 - 8.4.1\right)$$

$$R = 0.7\times555\times12\times\left(250/1.732\right)/1.1/1000 = 612\,kN$$

v)   **Cleat**
     Same as above; Provide 2 – ISA 100 100 12
vi) **Fillet Weld Sizing**
     Refer Figure 3.2.

Resultant force on the weld, $fr = 0.797\,kN/mm.$

Provide fillet weld, $\omega_1 = 6\,mm.$

$$\text{Permissible stress in weld}, fwd = \dfrac{\dfrac{fu}{\sqrt{3}}}{\gamma_{m0}} = 249\,MPa\left(IS\,800:2007\,10.5.7\right)$$

Strength of fillet weld connecting cleat angles to the web of supporting beam at field

$$= 0.8\times0.7\times6\times249/1000 = 0.837\,kN/mm > fr; Safe.$$

Refer Figure 3.3 above,

$$fr = 658\,kN\,/\,mm. \text{ Provide fillet weld}, \omega_2 = 6\,mm.$$

Strength of fillet weld connecting cleat angles to the web of supported beam at shop:

$$= 0.7.\omega_2.fwd\,N\,/\,mm = 0.7 \times 6 \times 249$$
$$= 1046\,kN\,/\,mm > fr\left(= 658kN\,/\,mm\right), \text{hence, Safe.}$$

3.1.3 WEB FRAMING CLEAT ANGLE – AISC (ASD)

a) **Sketch**
Same as Figure 3.1: Shear connection with angle cleats, above.
b) **Material**

TABLE 3.2
Properties of Structural Steel Material – AISC

ASTM A36

Yield stress, fy	F_{nBM}	36	ksi	250	MPa
Tensile stress	F_{uBM}	65	ksi	450	MPa
Elasticity	Es	29000	ksi	200000	MPa
Electrode Fe70		70	ksi		
Nominal stress	F_{nw}	42	ksi	292	MPa

Note: Nominal stress = 0.6 Fexx.

---

c) **Member Descriptions**

Supported member:	B1	W24 × 84	D = 24.1 inch.
Supporting member:	B2	W24 × 84	D = 24.1 inch.

d) **Design Load**
End shear, R = 70% of web shear capacity. (Assumed for Joint design)

$$\text{Effective depth of web}, D - h1 = 23.04\,inch, h1 = k = 1.06\,inch$$

$$\text{Web thickness}, tw = 0.47\,inch, h2 = h1 + 0.75 = 1.81\,inch$$

$$\text{Shear stress}, \tau sh = 0.6\,fy = 0.6 \times 36 = 21.6\,ksi$$

$$\text{Web shear Capacity} = Rn = 0.6\,fy.Agv = 0.6\,fy\left(D - h1\right).tw\left(AISC\,J\,4 - 3\right)$$

$$R = \text{Design endshear} = 0.7 \times \left(D - h1\right) \times tw \times \frac{0.6\,fy}{\Omega}, \text{where } \Omega = 1.67$$

$$= 0.7 \times 23.04 \times 0.47 \times 0.6 \times 36 / 1.67 = \mathbf{98}\,\text{kips}$$

e) Cleat Sizing

$$\text{Provide web cleat angles } 2 - L\,4 \times 4 \times 1/2$$

$$\text{Length of cleat}, Lv = 22\,\text{inch}$$

$$\text{Leg thickness}, Lt = 1/2\,\text{inch}$$

$$\text{Leg size}, Lh = 4\,\text{inch}$$

$$\text{Shear capacity of cleats} = 2 \times 22 \times 0.5 \times 21.6$$
$$= 475\,\text{kips} > R : \text{Okay.}\left(\text{AISC G2.1}\right)$$

f) Fillet Weld Sizing
Field weld to Supporting Beam, ω_1:
Please refer Figure 3.2: Welding of cleat on supporting beam web, above.

$$R = 98\,\text{kips}; fh = \left(9 \times 98 \times 4\right)/\left(5 \times 22^2\right) = 1.458\,\text{kip / inch}$$

$$Lh = 4\,\text{inch}; fv = 98/\left(2 \times 22\right) = 2.23\,\text{kip / inch}$$

$$Lv = 22\,\text{inch}; fr = \sqrt{\left(1.458^2 + 2.23^2\right)} = 2.66\,\text{kip / inch}$$

$$\text{Provide fillet weld of size}, \omega_1 = 3/8\,\text{inch}$$

$$\frac{Lv}{\omega_1} = 58.67 < 100; \text{effective length}$$
$$= \text{actual length}, Lv\left(\text{AISC} - \text{Table J2} - 2b\right)$$

$$\text{Permissible stress in fillet weld}, F_{nw} = 42\,\text{ksi};$$
$$\text{Electrode Fe} = 70\left(\text{AISC} - \text{Table J2} - 6\right)$$

(Nominal stress = 0.6 Fexx)
Strength of fillet weld connecting cleat angles to the web of supporting beam at field:

$$= 0.8 \times 0.7\omega_1\, F_{nw} / \Omega\,\text{kips / inch}\left(20\%\text{stress reduced for site weld}\right),$$

where $\Omega = 2$ (AISC Table J2.5)

$$= 0.8 \times 0.7 \times 0.375 \times 42 / 2 = 4.41\,\text{kips / inch} > fr; \text{Safe.}$$

Shop weld of cleat to supported beam, ω_2:

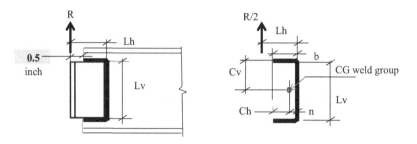

FIGURE 3.4   Shop welding of cleat on supported beam web.

In Figure 3.4,

$$n = b^2 / (2b + Lv)$$

$$Ch = Lv - n - 0.5$$

$$Cv = Lv / 2$$

$$b = Lh - 0.5$$

$$Jw = \left[(2b + Lv)^3 / 12 - b^2 (b + Lv)^2 / (2b + Lv) \right]$$

Twisting (horizontal)

$$fh = T Cv / Jw = \left[R(Lh - n)Cv / 2 Jw \right]$$

Twisting (vertical)

$$fv_1 = T Ch / Jw = \left[R(Lh - n)Ch / 2.Jw \right]$$

Shear (vertical)

$$fv_2 = \left[0.5 R / (2b + Lv) \right]$$

$$\mathbf{fr} = \sqrt{ \left[\mathbf{fh}^2 + (\mathbf{fv_1} + \mathbf{fv_2})^2 \right] }$$

In this example,

$$b = 3.5 \, \text{inch} \, (4 - 0.5); R = 98 \, \text{kips}$$

$$Lv = 22 \, \text{inch}$$

$$Lh = 4 \, inch$$

$$n = 0.42 \, inch \left(3.5^2 / \left(2 \times 3.5 + 22\right)\right)$$

$$Ch = 3.08 \, inch \left(4 - 0.42 - 0.5\right)$$

$$Cv = 11 \, inch \left(22 / 2\right)$$

$$Jw = 2032 - 275 = 1757 \, inch^3$$

$$fh = 1.098 \, kip / inch; 98 \times \left(4 - 0.42\right) \times 11 / \left(2 \times 1757\right)$$

$$fv1 = 0.308 \, kip / inch; 98 \times \left(4 - 0.42\right) \times 3.08 / \left(2 \times 1757\right)$$

$$fv2 = 1.69 \, kips / inch; 0.5 \times 98 / \left(2 \times 3.5 + 22\right)$$

$$fr = \sqrt{\left[1.098^2 + \left(0.308 + 1.69\right)^2\right]} = \mathbf{2.28} \, kips / inch$$

Provide fillet weld of size, $\omega_2 = 3 / 8 \, inch$.

Strength of fillet weld connecting cleat angles to the web of supporting beam at shop:

$$= 0.7 \omega_2 F_{nw} / \Omega \, kips / inch,$$

where $\Omega = 2$ (AISC Table J2.5)

$$= 0.7 \times 0.375 \times 42 / 2 = 5.51 \, kips / inch > fr; Safe.$$

3.1.4   WEB FRAMING CLEAT ANGLE – AISC (LRFD)

a) **Sketch** – See Figure 3.1
b) **Materials** – Same as in Table 3.2
c) **Member Description**

Supported member:	B1	W24 × 84	D = 24.1 inch.
Supporting member:	B2	W24 × 84	D = 24.1 inch.

d) **Design Load, R**
End Shear, $R = 70\%$ of web shear capacity. (Assumed for Joint design)

$$R = Design \, End \, shear = 0.7 \times \left(D - h1\right) \times tw \times 0.6 \, fy \times \phi,$$

where $\phi = 0.9$

$$= 0.7 \times 23.04 \times 0.47 \times 0.6 \times 36 \times 0.9 = \mathbf{147} \, kips$$

e) Cleat Sizing

Provide web cleat angles $2 - L4 \times 4 \times \frac{1}{2}$

Shear capacity of cleats $= 2 \times 22 \times 0.5 \times 21.6 = 475$ kips $>$ R; Okay.

f) Fillet Weld Sizing

Field weld to Supporting beam, ω_1,
See Figure 3.2.

Resultant force on the weld, $fr = \sqrt{\left(fh^2 + fv^2\right)}$

$R = 147$ kips; $fh = (9 \times 147 \times 4) / (5 \times 22^2) = 2.187$ kip / inch

$Lh = 4$ inch; $fv = 147 / (2 \times 22) = 3.34$ kip / inch

$Lv = 22$ inch; $fr = \sqrt{(2.187^2 + 3.34^2)} = 3.99$ kip / inch

Provide fillet weld of size, $\omega_1 = 1/4$ inch

Permissible stress in fillet weld, $F_{nw} = 42$ ksi; Electrode Fe $= 70$

Strength of fillet weld connecting cleat angles to the web of supporting beam at field:

$= 0.8 \times 0.7\omega_1 \, F_{nw} \times \phi$ kips / inch (20% stress reduced for site weld),

where, $\phi = 0.75$ (AISC Table J2.5)

$= 0.8 \times 0.7 \times 0.25 \times 42 \times 0.75 = 4.41$ kips / inch $> fr$; Safe.

Shop weld of cleat to supported beam, ω_2:
Refer Figure 3.4.

$R = 147$ kips

$fh = 1.647$ kip / inch, $147 \times (4 - 0.42) \times 11 / (2 \times 1757)$

$fv_1 = 0.461$ kips / inch, $147 \times (4 - 0.42) \times 3.08 / (2 \times 1757)$

$fv_2 = 2.534$ kips / inch, $0.5 \times 147 / (2 \times 3.5 + 22)$

$fr = \sqrt{\left[1.647^2 + (0.461 + 2.534)^2\right]} = 3.42$ kips / inch

Provide fillet weld of size, $\omega_2 = 1/4$ inch.

Strength of fillet weld connecting cleat angles to the web of supported beam at shop:

$$= 0.7\omega_2 F_{nw}.\phi \text{ kips / inch,}$$

where $\phi = 0.75$ (AISC Table J2.5)

$$= 0.7 \times 0.25 \times 42 \times 0.75 = 5.51 \text{ kips / inch} > \text{fr; Safe.}$$

3.1.5   END CONNECTING FACE PLATE – IS 800 (WORKING STRESS)

a) Sketch

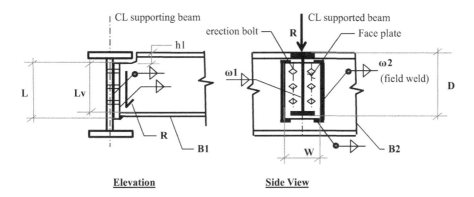

FIGURE 3.5   Shear connection with face plate – welded.

This type of web face plate connection is not flexible like web cleat joints, where the cleat angles tend to twist or rotate under the application of reaction force. In a face plate type joint, the connections are rigid and there is no spreading action, which would be resisted by welds. Hence, the vertical fillet welds should be designed for just the vertical reaction, R. The Leg size of this fillet weld must be equal to the web thickness, based upon the standard allowable, if it is to match the allowable strength of this web section in shear as well as tension.

b) Materials
Refer Table 3.1

c) Member Description

| Supported member: | B1 | ISMB 450 | $D = 450$ mm |
| Supporting member: | B2 | ISMB 450 | $D = 450$ mm |

d) Design Load
End Shear, R = 70% of web shear capacity. (assumed)

Effective depth of web, $D - h1 = 414$ mm, $h1 = 36$ mm

Web thickness, tw = 9.4 mm.

Shear stress, τsh = 0.4 fy = 0.4 × 250 = 100 Mpa. (IS 800 : 2007 – 11.4.2)

$$R = \text{End shear} = 0.7 \times (D - h1) \times tw \times 0.4\,fy\,/\,1000$$

$$= 0.7 \times 414 \times 9.4 \times 0.4 \times 250\,/\,1000 = \mathbf{272}\ kN.$$

e) Face Plate Sizing

$$Lv = R / (tw.0.4fy) = 272000 / (9.4 \times 0.4 \times 250) = 290\ mm\,(rounded).$$

$$L = Lv + \text{flange thickness} + 12\ mm$$

$$L = 290 + 17.4 + 12 = 320\ mm;\ \text{Provide}\ L = 424\ mm.$$

W = width of the plate = width of the flange of the supported beam.

$$W = 150\ mm.$$

Thickness of plate = Thickness of web of the supported beam.

$$tp = t = 9.4\ mm,\ \text{say}\ 10\ mm.$$

f) Fillet Weld Sizing
Electrode: EX40XX; fy = 330 MPa
Permissible stress in fillet weld, fw = 132 MPa (Shop weld, 0.4 fy)

$$\omega_1 = \text{fillet weld at shop} = 10\ mm\,(= \text{web thickness}).$$

Strength of fillet welding connecting supported beam web to face plate = F1.

$$F1 = 2Lv.0.7.\omega_1.fw = 2 \times 290 \times 0.7 \times 10 \times 132\,/\,1000 = 536\ kN > R;\ Safe.$$

Permissible stress in fillet weld, fw′ = 106 MPa (Site weld = 80% of shop weld)

$$\omega_2 = \text{fillet weld at site} = 10\ mm\,(= \text{web thickness}).$$

Strength of fillet welding connecting the face plate to the web of supporting beam = F2

$$F2 = 2L.0.7.\omega_2.fw' = 2 \times 320 \times 0.7 \times 10 \times \frac{105}{1000} = 473\ kN > R$$

3.1.6   END CONNECTING FACE PLATE – IS 800 (LIMIT STATE)

a) **Sketch**

   Refer Figure 3.5

b) **Materials**

   Refer Table 3.1

c) **Member Description**

Supported member:	B1	ISMB 450	$D = 450$ mm
Supporting member:	B2	ISMB 450	$D = 450$ mm

d) **Design Load**

   End Shear, R = 70% of web shear capacity. (assumed)

$$\text{Effective depth of web, } D - h1 = 414 \, mm, h1 = 36 \, mm$$

$$\text{Web thickness, tw} = 9.4 \, mm.$$

$$\text{Shear stress, } \tau sh = fy \, / \, \sqrt{3} = 250 / 1.732 = 144 Mpa. \left(IS \, 800 : 2007 - 8.4.1 \right)$$

$$R = \text{End shear} = 0.7 \times \left(D - h1 \right) \times tw \times \left(fy \, / \, \sqrt{3} \right) / \gamma_{m0} / 1000$$

$$= 0.7 \times 414 \times 9.4 \times 144 / 1.1 / 1000 = \mathbf{357} \, kN.$$

e) **Face Plate Sizing**

$$Lv = R / \left(tw.fy \, / \, 1.732 \right) = 357000 / \left(9.4 \times 144 \right) = 265 \, mm$$

$$\text{Length, L} = Lv + \text{flange thickness} + 12 \, mm = 265 + 17.4 + 12 = 295 \, mm,$$

$$\text{Provide L} = 420 \, mm$$

$$\text{Width, W} = 150 \, mm, \text{Thickness, tp} = 10 \, mm$$

f) **Fillet Weld Sizing**

$$\text{Electrode: EX40XX, fy} = 330 \, MPa, fu = 540 \, MPa$$

$$\text{Permissible stress in fillet weld, fw} = 132 \, MPa \, \text{Shop weld} (= fu / \sqrt{3} / \gamma_{mw})$$

$$\omega_1 = \text{fillet weld at shop} = 6 \, mm$$

$$\text{Strength of fillet welding connecting supported beam web to face plate} = F1$$

$$F1 = 2 \times 265 \times 0.7 \times 6 \times \frac{249}{1000} = 554 \, kN > R; \text{Safe}.$$

Permissible stress in fillet weld, $fw' = 199$ MPa (Site weld = 80% of shop weld)

$$\omega_2 = \text{fillet weld at site} = 6\,\text{mm}\,(= \text{web thickness}).$$

Strength of fillet welding connecting the face plate to the web of supporting beam = F2

$$F2 = 2L.0.7.\omega_2.fw' = 2 \times 295 \times 0.7 \times 6 \times 199\,/\,1000 = 494\,\text{kN} > R;\text{Safe}.$$

3.1.7 END CONNECTING FACE PLATE – AISC (ASD)

a) **Sketch**
   Refer Figure 3.5.
b) **Material**
   Refer Table 3.2.
c) **Member Description**

Supported member:	B1	$W12 \times 26$	$D = 12.22$ inch.
Supporting member:	B2	$W12 \times 26$	$D = 12.22$ inch.

d) **Design Load, R**
   End Shear, $R = 70\%$ of web shear capacity. (assumed for Joint design)

$$\text{Effective depth of web}, D - h1 = 11.16\,\text{inch}; h1 = k = 1.06\,\text{inch}$$

$$\text{Web thickness}, tw = 0.23\,\text{inch}; \text{Flange thickness}, tf = 0.38\,\text{inch}$$

$$\text{Shear stress}, \tau sh = 0.6\,fy = 0.6 \times 36 = 21.6\,\text{ksi}$$

$$\textit{Web shear capacity} = Rn = 0.6\,fy.Agv = 0.6\,fy\,(D - h1).tw\,(\text{AISC J 4-3})$$

$$R = \text{Design end shear} = 0.7 \times (D - h1) \times tw \times \frac{0.6\,fy}{\Omega}, \text{where } \Omega = 1.67$$

$$= 0.7 \times 11.16 \times 0.23 \times 0.6 \times 36\,/\,1.67 = \textbf{23}\,\text{kips}.$$

e) **Face Plate Sizing**

$$\text{Lv required} = R\,/\,(tw.0.6\,fy) = 23\,/\,(0.23 \times 0.6 \times 36) = 5\,\text{inch}.$$

$$\text{L provided} = D - h1 + 0.5\,\text{inch}.$$

$$L = 12.2 - 1.06 + 0.5 = 12\,\text{inch} > \text{Lv reqd}; \text{Okay}.$$

W = width of the plate = width of the flange of the supported beam.
W = 6.49 inch. Provide 7.5 inch wide plate.
Thickness of the plate ≥ Thickness of web of supported beam.

$$tp = tw = 5/16 \text{ inch (provided)}$$

f) Fillet Weld Sizing

$$\text{Electrode Fe 60 fy} = 60 \text{ ksi}; \ \Omega = 2$$

Permissible stress in fillet weld, fw = 18 ksi (shop weld) [fw = 0.6 fy/Ω]

$$\omega_1 = \text{fillet weld size} = 1/4 \ \text{inch}$$

Strength of fillet welding connecting supported beam web to face plate, F1

$$F1 = 3(D - h1 - tf).0.7\omega_1 fw$$
$$= 2 \times (12.22 - 1.06 - 0.38) \times 0.7 \times 0.25 \times 18 = 68 \text{ kips.}$$

F1 > R; Safe.
Permissible stress in fillet weld (site), fw' = 14 ksi (shop weld) [80% of shop weld]

$$\omega_2 = \text{fillet weld size} = 1/4 \text{ inch (site weld)}$$

Strength of fillet welding connecting the face plate to the web of supporting beam, F2

$$F2 = 2L.0.7 \ \omega_2 \ fw' = 2 \times 11.66 \times 0.7 \times 0.25 \times 14 = 57 \text{ kips} > R \text{ safe.}$$

3.1.8   END CONNECTING FACE PLATE – AISC (LRFD)

a) Sketch
Refer Figure 3.5.
b) Material
Refer Table 3.2.
c) Member Description

Supported member:	B1	W12 × 26	D = 12.22 inch.
Supporting member:	B2	W12 × 26	D = 12.22 inch.

d) Design Load, R
End Shear, R = 70% of web shear capacity. (Assumed for Joint design)

$$\text{Effective depth of web,} D - h1 = 11.16 \text{ inch;} h1 = k = 1.06 \text{ inch}$$

Web thickness, tw = 0.23 inch; Flange thickness, tf = 0.38 inch

Shear stress, τsh = 0.6fy = 0.6 × 36 = 21.6 ksi (AISC J 4 – 3)

R = Design End shear = 0.7 × (D – h1) × tw × 0.6fy × ϕ
where ϕ = 0.9

$\quad$ = 0.7 × 11.16 × 0.23 × 0.6 × 36 × 0.9 = **35** kips.

e) Face Plate Sizing

Lv required = R / (tw.0.6fy) = 35 / (0.23 × 0.6 × 36) = 7 inch.

L provided = D – h1 + 0.5 inch..

L = 12.2 – 1.06 + 0.5 = 12 inch > Lv reqd; Okay.

W = width of the plate = width of the flange of the supported beam.
W = 6.49 inch., Provide 7.5 inch-wide plate.
Thickness of the plate ≥ Thickness of web of supported beam.

tp = tw = 5 / 16 inch (provided)

f) Fillet Weld Sizing

Electrode Fe 60, fy = 60 ksi, ϕ = 0.75
Permissible stress in fillet weld, fw = 27 ksi (shop weld) [fw = ϕ. 0.6 fy]

ω_1 = fillet weld size = 1 / 4 inch

Strength of fillet welding connecting supported beam web to face plate, F1

F1 = 3 (D – h1 – tf).0.7$\omega_1$fw
$\quad$ = 2 × (12.22 – 1.06 – 0.38) × 0.7 × 0.25 × 27 = 102 kips.

F1 > R; Safe

Permissible stress in fillet weld (site), fw' = 22 ksi (shop weld) [80% of shop weld]

ω_2 = fillet weld size = 1 / 4 inch (site weld)

Strength of fillet welding connecting the face plate to the web of supporting beam, F2

F2 = 2 L.0.7 ω_2 fw' = 2 × 11.66 × 0.7 × 0.25 × 22 = 90 kips > R safe.

3.2   BEAM TO COLUMN – MOMENT CONNECTION (MC)

a) Reference
1. IS 800:2007

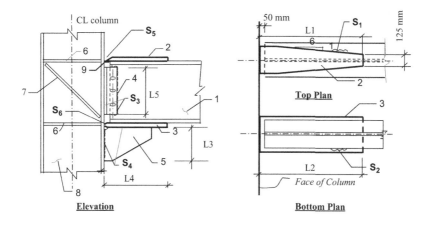

Items:

1 Plate Girder / Rolled steel Joist
2 Top cover plate
3 Bottom cover plate
4 Web cleat [wired on erection bolt; slotted hole]
5 Bracket/ Vertical stiffener
6 Column stiffener / Horz ribs.
7 Column stiffener / Diagonal rib.
8 Column
9 Back strip

Weld details:

S1 Fillet weld at top cover plate
(Site weld; Single V butt on back strip).
S2 Fillet weld at bot. flange plate (Site weld).
S3 Fillet weld at web cleat (Site weld).
S4 Fillet weld at web stiffener.
S5 Full penetration butt weld (Site weld).
S6 Full penetration butt weld (Double V).

FIGURE 3.6   Moment connection joint – welded.

b) Materials

TABLE 3.3
Properties of Structural Materials – MC Joints

Yield stress	Fy	36	ksi	250	MPa
Tensile stress	Fu	65	ksi	450	MPa
Plate bending stresses	σ_{bt}		tension	150	MPa
	σ_{bc}		comp	165	MPa
Fillet weld				110	MPa
Electrode Fe 70		70	ksi		
Nominal stress	Fnw	42	ksi	290	MPa
Elasticity	E	29000		200000	Mpa

c) **Sketch**

Refer to Figure 3.6.

d) **Joint Design Forces and Member Sections**

Design Moment, M = 5440 kNm;	Axial Load, H = 75 kN.
Design Shear, F = 1230 kN	*(50% of Web shear capacity)*

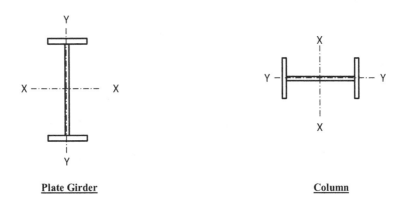

Plate Girder	Column

FIGURE 3.7   Typical Cross sections.

Beam/Plate Girder	**Columns**
Flange Plate – 600 mm × 32 mm	Flange Plate – 600 mm × 28 mm
Web Plate – 1536 mm × 16 mm	Web Plate – 1144 mm × 28 mm
Unsupported Length of Beam	Unsupported Length of Column
Span Lx = 12 m; Ly = 2.5 m	Height Lx = 12 m; Ly = 6 m

e) **Top Cover Plate – Item 1**

Depth of girder, D = 1.6 m; Chord force, T = M/D = 3400 kN

$$\text{Area required, Arqd} = T / \sigma_{bc} = 3400 \times 1000 / 165 = 20606 \ \text{mm}^2$$

$$\text{Allowable bending compression} = \sigma_{bc} = 165 \, \text{MPa}$$

Yield stress, fy = 240 MPa (for 40 mm > plate thickness > 20mm)
Size of Top plate: L1 = 1600 mm, B1 = 530 mm, T1 = 45 mm

Size of Fillet weld: S1 = 20 mm (Site weld).

$$Cross-section\ area\ of\ cover\ plate = B1 \times T1 = 530 \times 45$$
$$= 23850 mm^2 > Arqd + 5\%; Okay$$

(Area of cover plate should not be less than 5% in excess of the area of girder flange.)

Total length of fillet weld required = T / Strength of weld per mm.

Hence,

$$L_{weld} = 3400 \times 1000 / (0.7 \times 20 \times 110 \times 0.8)$$
$$= 2760\ mm\ (Reduction\ factor\ for\ site\ weld = 0.8)$$

The cover plate will have two runs of fillet
weld S1 and end run of 125 mm.

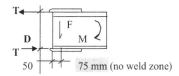

Elevation

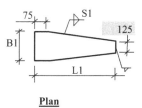

Plan

FIGURE 3.8   Top cover plate.

So, the required length of cover plate

$$= (Lweld - 125\ mm) \times 0.5 + no\ weld\ zone\ (75\ mm) = 1393\ mm.$$

Length of cover plate provided, L1 = 1600 mm > 1393 mm; Safe.

f) Bottom Cover Plate – Item 3
Width of bottom flange of Beam = 600 mm
   Size of bottom plate: L2 = 1600 mm, B2 = 710 mm, T2 = 32 mm
   Size of fillet weld: S2 = 16 mm (Site weld).

Total length of fillet weld S2 will be = T/ strength of weld per mm.

$$Lweld = 3400 \times 1000 / (0.7 \times 16 \times 110 \times 0.8) = 3450 \text{ mm}.$$

There will be two runs of fillet weld S2 + end run (width of bottom flange of MC beam).

So, reqd. length of cover plate = (Lweld – width of bottom flange) 0.5 + No weld zone.

$$= (3450 - 600) \times 0.5 + 75 = 1500 \text{ mm}.$$

Length of cover plate provided, L2 = 1600 mm > 1500 mm; Okay.

g) Column Web Stiffener

Checking shear stress in knee (shown hatched).
   (Refer Steel designer's manual.)
   Max shear stress, v

$$= \left[Mc / 4.a.b.t \right].\left[1 + a^2.t / (3.Za) + b^2.t / (3.Zb) \right]$$

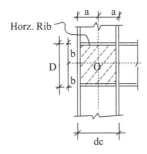

FIGURE 3.9   Column web stiffeners.

Mc = M−V.a −H.b, tweb = plate thickness
Za = Sectional modulus of the knee along horizontal axis. = I column/a
Zb = Sectional modulus of the knee along vertical axis. = I beam/b
M = Beam moment; V = End shear; H = Axial load in beam
dc = 1.2 m; M = 5440 kNm; I column = 0.015 m4
D = 1.6 m; V = 1230 kN; I beam = 0.028 m4
$t_{web} = 0.028\,m_,$ H = 75 kN

$$a = 0.6 \text{ m}; b = 0.8 \text{ m}; Mc = 4642 \text{ kNm}; Mc / 4abt = 86347$$

$$Za = 0.02506 \text{ m}^3; \ Zb = 0.03555 \text{ m}^3$$

$$a^2.t/(3.za) = 0.134; \ b^2.t/(3.zb) = 0.168$$

$$v = 86347 \times (1 + 0.134 + 0.168)/1000 = 112 \, \text{Mpa} > 100 \, \text{MPa}; \text{Safe}$$

$$\text{Allowable shear stress}, \tau_{va} = 0.4 \times 250 = 100 \, \text{MPa}.$$

Hence, additional stiffening of web will not be required.

h) Horizontal Stiffener/Rib – Item 6

$$T = 3400 \, \text{kN} \quad (= M/D)$$

$$\tau_{va} = 100 \, \text{MPa} (= 0.4 \text{fy})$$

Col. web thickness, w2 = 28 mm
Depth of column, dc = 1200 mm
D = 1600 mm
Now, horizontal shear resisted by web,

$$tc = \tau va.dc.w2 = 3360 \, \text{kN}$$

$$Fh = T - tc = 40 \, \text{kN}.$$

Required area of horizontal stiffeners, **Arqd** = Fh / 0.6 fy = 267 mm^2.

Provide 2 Rib plates – wide = 250 mm; thickness = 16 mm;

$$\text{Area} = 2 \times 250 \times 16 = 8000 \, \text{mm}^2 > A \, \text{required} (267 \, \text{mm}^2); \text{Okay}.$$

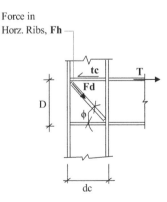

FIGURE 3.10 Force in stiffener.

i) Diagonal Ribs (Item 7)/Doubler Plates

$$\mathbf{Fd} = Fh / \cos \phi = 40 / 0.6 = 67 \, kN$$

Required area of Diagonal stiffener

$$= Fd / 0.6fy = 444 \, mm^2$$

Provide 2 Rib plates: 100 mm × 12 mm

Total area provided $\left(2 \text{ diagonal ribs} + 20.\text{tweb}^2\right) = 18080 \, mm^2 >$ required area.

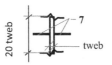

FIGURE 3.11 Cross section of item 7.

Alternate: Doubler plates

Web doubler plate: Not used

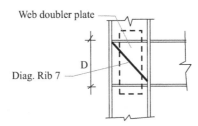

FIGURE 3.12 Web stiffener

j) Web Cleats and Bracket – Item 4 and 5

Design end shear force, **F = 1230 kN**

The end shear F will be transferred by two web cleats (item 4) and the vertical stiffener (item 5)

Web cleats: 2 ISA 100 100 10; height = 1300 mm.

Vertical stiffener plate: Length = 600 mm; Height = 250 mm; Thickness = 12 mm

Design of weld

Height of web cleats, L5 = 1300 mm

Height of web stiffener, L3 = 250 mm

Size of Fillet weld, S3 = 8 mm (site weld); Strength = 0.7 × 8 × 110 × 0.8 = 493 N/mm

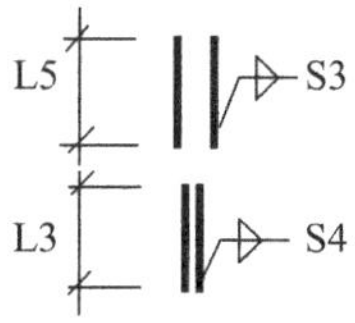

FIGURE 3.13 Weld run Shear.

Size of Fillet weld, S4 = 6 mm (shop weld); Strength = $0.7 \times 6 \times 110 = 462$ N/mm

Shear Capacity of weld run = $(2 \times 1300 \times 493 + 2 \times 250 \times 4620)/1000 = 1512\,kN$

Shear capacity of cleats and stiffener

Thickness of web cleats, T3 = 10 mm; Thickness of web stiffener, T4 = 12 mm

Shear Capacity of web cleats = $2.L5.\tau_{va}$
$$= 2 \times 1300 \times 10 \times 0.4 \times 250 / 1000 = 2600\,kN$$

Shear Capacity of web stiff. = $L3.\tau_{va} = 250 \times 12 \times 0.4 \times 250 / 1000 = 300kN$

Total shear capacity = $2600 + 300 = 2900\,kN > F; Okay.$

TABLE 3.4
Summary of Calculation

Beam	2 Flng	600	mm	32	mm	1 Web	1536	mm	16	mm
Column	2 Flng	600	mm	28	mm	1 Web	1144	mm	28	mm

Connection Materials

Top Cover Plate			**Bottom Plate**				**Weld Sizes**			
L1	B1	T1	L2	B2	T2		S1	S2	S3	S4
1600	530	45	1600	710	32		20	16	8	6

Cleat Angles						**Bracket**		**Horz Ribs on Col Web**		
nos	leg1	leg2	thick	height	height	length	thick	nos	wide	thick
		OSL	T3	L5	L3	L4	T4			T5
	mm	mm	mm	mm	mm	mm	mm		mm	mm
2	100	100	10	1300	250	600	12	2	250	16

Additional Column Web Stiffener

Diagonal Ribs			**Doubler Plate**		
nos	wide	thick	nos	wide	thick
T6			T7		
	mm	mm		mm	mm
2	100	12			

1. OSL means outstanding leg.
2. All welds are 6 mm fillet weld U.N.O.
3. Top and bottom cover plates will have full penetration butt weld.

3.3   TRUSS MEMBER JOINTS

3.3.1   Description

Truss member joint is a detail showing all the members meeting at a point or node
and joined together by a common gusset plate. The members are connected to the
gusset plate by bolts or weld to transfer the tension or compression force at the joint.
The shape of gusset is developed according to the array of bolts or weld length necessary for transferring forces from each member end.

The member forces are usually given in the design drawing for joint design; otherwise, the designer should use full strength of the member section (tension or compression) to determine number of bolts or length and size of weld. Do not use bolts
and weld in combination to resist the member end force.

In this section, we will provide a method of calculation to find out the full strength
of a member section and the corresponding welding requirement.

3.3.2   Sketches of Truss Member Joints

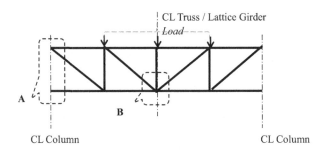

FIGURE 3.14   Elevation – Truss/Lattice girder.

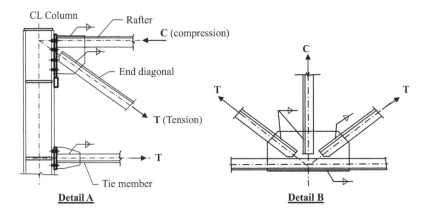

FIGURE 3.15   Truss member joints.

3.3.3   Single Angle Section – AISC (ASD)

The truss members are generally made of double angles, single angle, channels, and I-section base on member load and span. The lateral stability of member section is also seen while calculating the member forces for joint design of a truss member.

a) **Materials**
   Refer Table 3.2
b) **Design Parameters**

$$\text{Size of angle} = \mathbf{L\,4\times4\times1/2}$$

$$\text{Fillet weld,}\,\omega = 1/4\,\text{inch}$$

$$\text{Length / side,}\,Lw = 12\,\text{inch}$$

$$\text{Gusset thickness,}\,t = 5/16\,\text{inch}$$

$$\text{Hole dia. for erection bolt,}\,d = 0.61\,\text{inch}\,(M14)$$

$$\text{No. of bolt holes} = 3$$

$$b_L = bs = 4\,\text{inch.;}\,t = 0.5\,\text{inch}$$

$$Ag = 3.75\,\text{in}^2;\,An = 3.45\,\text{in}^2$$

$$\text{Fillet weld size,}\,\omega = 6\,\text{mm}\ \left[=4/16\,\text{inch} = D/16\,\text{inch};\,D = 4\right]$$

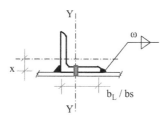

FIGURE 3.16   Single angle welded to gusset.

c) **To Determine Tensile Strength, T** (AISC D2-1)
   Tensile strength, T shall be the lower value of tensile yielding in gross section (Pn1) and tensile rapture in net section (Pn2).
   i)   Tensile yielding, Pn1 = FyAg/Ω = 36 × 3.75/1.67 = 80.84 kips

$$\Omega = 1.67\ \text{ASD}$$

ii) Tensile rupture, $Pn2 = Fu. Ae/\Omega = Fu An. U/\Omega = 65 \times 3.45 \times 0.9/2 = 101$ kips.

$$\Omega = 2\,ASD; Ae = An\,U, \text{where } U = \text{shear lag factor per Table D3.1} (AISC)$$

$$U = 1 - x / L; L = 305\,mm \,(L = \text{length of connection})$$

$$X = Cyy = 30\,mm; U = 0.8 \text{ with 4 fastener Case 8 per Table D3.1} (AISC)$$

$$U = 1 - (30 / 305) = 0.9 \text{ with Case 2 per Table D 3.1} (AISC)$$

Allowable tension $T = 80.84$ kips

d) To Determine Compressive Strength, C

Member : $\mathbf{L\,4 \times 4 \times 1 / 2}$

Member slenderness check : $b / t <= 0.45 \sqrt{(E / Fy)} (AISC - Table\,B4.1a)$

$$0.45 \sqrt{(E / Fy)} = 0.45 \sqrt{(29000 / 36)} = 12.8$$

$$b / t = 4 / 0.5 = 8 \text{ Non Slender}$$

Allowable compressive strength, $C = Pn3 / \Omega = 124 / 1.67 = 74$ kips; $\Omega = 1.67\,(ASD)$

Nominal compressive strength, $Pn = Fcr.Ag = 33 \times 3.75 = 124$ kips. $(AISC\,E3-1)$

Length of member, $Le = 36$ inch (assumed); $K = 0.85; r_{min} = 0.776$ inch.

$$\lambda = KL / r = 0.85 \times 36 / 0.776 = 39$$

$$4.71 \sqrt{(E / Fy)} = 4.71 \times \sqrt{(29000 / 36)} = 134$$

Elastic buckling stress, $Fe = \pi^2 E / \lambda^2 \text{ ksi} = 3.14^2 \times 29000 / 39^2 = 188$ ksi

$$KL / r <= 4.71 \sqrt{E / Fy}; Fcr = \left[0.658^{(Fy/Fe)} \right] Fy$$

$$Fcr = \left[0.658^{(36/188)} \right] \times 36 = 33\,ksi$$

Allowable compression, $C = 74$ kips

e) To Determine Block Shear Resistance, V

$$V = Rn / \Omega;$$

$$Rn = 0.6\,Fu\,Anv + Ubs\,Fu\,Ant \quad <= 0.6\,Fy\,Agv + Ubs\,Fu\,Ant\left(AISC\,J4-5\right)$$

$$\Omega = 2\left(ASD\right)$$

where,

$$Agv = \text{Gross area subject to shear yielding}\left(2Lw.t\right) = 2\times12\times\left(5/16\right) = 7.5\,in^2$$

$$Anv = \text{net area subject to shear rupture }\left(Agv - \Sigma\,\text{hole area}\right)$$
$$= 7.5 - 3\times0.5\times0.61 = 6.59\,in^2$$

$$Ant = \text{net area subject to tension}\left(bs.t\right) = 4\times0.5 = 2\,in^2$$

Ubs = 1, where tension stress is uniform; 0.5, where tension stress is non-uniform.

The block shear failure mode must be checked around periphery of welded connection.

(AISC J4-3 -16.1-411)

$$\text{Now}, 0.6\,Fu\,Anv + Ubs\,Fu\,Ant = 0.6\times65\times6.59 + 1\times65\times2 = 387\,kips$$

$$0.6\,Fy\,Agv + Ubs\,Fu\,Ant = 0.6\times36\times7.5 + 1\times65\times2 = 292\,kips$$

$$Rn = 292\,kips. V = Rn / \Omega = 292 / 2 = 146\,kips$$

$$\text{Block shear strength}, V = 146\,kips. > T; \text{hence}, Okay.$$

Let us consider the maximum of tensile and compressive strength for a full strength joint design.

Full Strength, F = 80.84 kips.

Strength of fillet weld resisting full strength of member:

$$Lw = 12\,inch / side$$

$$\omega = 1/4\,inch$$

$$Fnw = 42\,ksi\left(shop\right)$$

$$\Omega = 2\left(ASD\right)$$

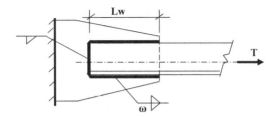

FIGURE 3.17   Plan view – welded truss member.

Strength of fillet weld connecting the single angle to the gusset plate:

$$= 0.7.\omega.\text{Fnw} \times 2 \ \text{Lw} / \Omega \ \text{kips}$$

$$= 0.7 \times 0.25 \times 42 \times (2 \times 12) / 2 = 88 \ \text{kips} > \text{F}; \text{Safe}.$$

To check thickness of gusset: (AISC 9-2)

$$t_{min} = 0.6 \, \text{Fexx} \left(\sqrt{2} / 2 \right) \left(D / 16 \right) / 0.6 \, \text{Fu},$$

where Fexx = 70ksi; Fu = 65ksi and

$$\text{weld size} = D / 16 \, \text{inch} \left(D = 4 \right)$$

$$= 0.6 \times 70 \times 0.7 \times (4 / 16) / (0.6 \times 65) = 0.188 \, \text{inch} = 1 / 5 \, \text{inch} \left(5 \, \text{mm} \right)$$

Gusset thickness is 5/16 inch (8 mm); hence, Okay.

3.3.4   DOUBLE ANGLE SECTION – AISC (ASD)

The double angle sections are commonly used in various types of roof trusses. The angles are placed back-to-back and separated by gussets at joints and pack plates along the length of members at specified intervals. The designs of joints with double angles are presented in the example below.

a) Material
   Refer Table 3.2
b) Design Parameters

$$\text{Member section} = 2 - \text{L} \, 4 \times 4 \times 1 / 2$$

$$\text{Fillet weld,} \, \omega = 1 / 4 \, \text{inch}$$

$$\text{Length} / \text{side}, \text{Lw} = 12 \, \text{inch}$$

$$\text{Gusset thickness}, t = 5/16 \text{ inch}$$

$$\text{Hole dia.for erection bolt}, d = 0.61 \text{ inch}\,(M14)$$

$$\text{No.of bolt holes} = 3$$

$$b_L = bs = 4 \text{ inch.} t = 0.5 \text{ inch}$$

$$Ag = 7.5 \text{ in}^2; An = 6.89 \text{ in}^2$$

$$\text{Fillet weld size}, \omega = 6 \text{ mm}\left[=4/16 \text{ inch} = D/16 \text{ inch}; D = 4\right]$$

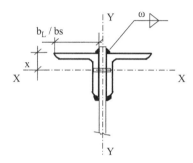

FIGURE 3.18 Double angle member joint.

c) To Determine Tensile Strength, T

Tensile strength, T shall be the lower value of tensile yielding in gross section (Pn1) and tensile rapture in net section (Pn2).

i) Tensile yielding, Pn1 = FyAg/Ω = 36 × 7.5/1.67 = 162 kips
Ω = 1.67 ASD

ii) Tensile rupture, Pn2 = Fu. Ae/Ω = Fu An. U/Ω = 65 × 6.9 × 0.9/2

$$= 202 \text{ kips.}$$

$$\Omega = 2 \text{ ASD}; Ae = An\, U, \text{where } U = \text{shear lag factor per Table D3.1}\,(\text{AISC})$$

$$U = 1 - x/L; L = 305 \text{ mm}\,(L = \text{length of connection})$$

$$X = Cyy = 30 \text{ mm}; U = 0.8 \text{ with 4 fastener Case 8 per Table D3.1}\,(\text{AISC})$$

$$U = 1 - (30/305) = 0.9 \text{ with Case 2 per Table D 3.1}\,(\text{AISC})$$

$$\text{Allowable tension T} = 162 \text{ kips}$$

d) To Determine Compressive Strength, C

$$\text{Member}: 2 - L4 \times 4 \times 1/2$$

$$\text{Member slenderness check}: b/t <= 0.45\sqrt{E/Fy} \left(\text{AISC} - \text{Table B4.1a}\right)$$

$$0.45\sqrt{(E/Fy)} = 0.45\sqrt{(29000/36)} = 13$$

$$b/t = 4/0.5 = 8 \text{ Non Slender}$$

$$\text{Allowable compressive strength}, C = Pn3/\Omega = 263/1.67 = 157 \text{ kips}; \text{gg}\Omega = 1.67$$

$$\text{Nominal compressive strength}, Pn = Fcr.Ag = 35 \times 7.5 = 263 \text{ kips}. \left(\text{AISC E3} - 1\right)$$

$$\text{Length of member}, Le = 36 \text{ inch} \left(\text{assumed}\right); K = 0.85; r_{min} = 1.21 \text{ inch}.$$

$$\lambda = KL/r = 0.85 \times 36/1.21 = 25$$

$$4.71\sqrt{(E/Fy)} = 4.71 \times \sqrt{(29000/36)} = 134$$

$$\text{Elastic buckling stress}, Fe = \pi^2 E/\lambda^2 \text{ ksi} = 3.14^2 \times 29000/25^2 = 457 \text{ ksi}$$

$$KL/r <= 4.71\sqrt{(E/Fy)} \quad Fcr = \left[0.658^{(Fy/Fe)}\right]Fy$$

$$Fcr = \left[0.658^{(36/457)}\right] \times 36 = 35 \text{ ksi}$$

$$\text{Allowable compression}, C = 157 \text{ kips}.$$

e) To Determine Block Shear Resistance, V

$$V = Rn/\Omega;$$

$$Rn = 0.6 Fu Anv + Ubs Fu Ant <= 0.6 Fy Agv + Ubs Fu Ant \left(\text{AISC J4} - 5\right)$$

$$\Omega = 2 \left(\text{ASD}\right)$$

where,

$$Agv = \text{Gross Area subject to shear yielding} \left(4Lw.t\right) = 4 \times 12 \times \left(5/16\right) = 15 \text{in}^2$$

$$Anv = \text{net area subject to shear rupture} \left(Agv - \Sigma \text{ hole area}\right)$$
$$= 15 - 3 \times 2 \times 0.5 \times 0.61 = 13.17 \text{in}^2$$

$$Ant = net\,area\,subject\,to\,tension\,(2\,bs.t) = 2\times4\times0.5 = 4\,in^2$$

Ubs = 1, where tension stress is uniform; 0.5, where tension stress is non-uniform.

$$Now, 0.6\,Fu\,Anv + Ubs\,Fu\,Ant = 0.6\times65\times13.17 + 1\times65\times4 = 774\,kips$$

The block shear failure mode must be checked around periphery of welded connection. (AISC J4-3 16.1-411)

$$0.6\,Fy\,Agv + Ubs\,Fu\,Ant = 0.6\times36\times15 + 1\times65\times4 = 584\,kips$$

$$Rn = 584\,kips. V = Rn\,/\,\Omega = 584\,/\,2 = 292\,kips$$

$$Block\,shear\,strength, V = 292\,kips. > T; hence\,Okay.;$$

Let us consider the maximum of tensile and compressive strength for a full strength joint design.

Full Strength, F = 162 kips.

Strength of fillet weld resisting full strength of member:
  Refer Figure 3.17

$$Lw = 12\,inch\,/\,side; \omega = 1\,/\,4inch; Fnw = 42\,ksi\,(shop)$$

$$\Omega = 2\,(ASD)$$

Strength of fillet weld connecting the single angle to the gusset plate:

$$= 0.7.\omega.Fnw \times 4Lw\,/\,\Omega\,kips$$

$$= 0.7\times0.25\times42\times(4\times12)/2 = 176\,kips > F; Safe.$$

To check thickness of gusset: (AISC 9-2)

$$t_{min} = 0.6\,Fexx(\sqrt{2}/2)(D/16)/0.6\,Fu,$$

where Fexx = 70ksi; Fu = 65ksi and
  weld size = D/16 inch (D = 4)

$$= 0.6\times70\times0.7\times(4/16)/(0.6\times65) = 0.188\,inch = 1/5\,inch\,(5\,mm)$$

Gusset thickness is 5/16 inch (8 mm); hence, Okay.

3.3.5   SINGLE ANGLE SECTION – IS 800 (WORKING STRESS)

a) Material:
Refer Table 3.1

b) Design Parameters:
Refer Figure 3.16
Member section = **ISA 100 100 10**

$$\text{Fillet weld}, \omega = 8\,\text{mm}$$

$$\text{Weld length / side.Lw} = 300\,\text{mm}$$

$$\text{Gusset thickness}, \text{tg} = 12\,\text{mm}$$

$$\text{Hole dia.for erection bolt}, d = 15.50\,\text{mm}$$

$$2\,\text{nos}\left(\text{M14}\right)$$

$$b_L = bs = 100\,\text{mm} \quad t = 10\,\text{mm} \quad Ag = 1903\,\text{mm}^2 \quad An = 1748\,\text{mm}^2$$

c) Tensile Strength of Member, Ts (IS 800:2007 -11.2)
Tensile strength, Ts shall be the lower value of tensile yielding in gross section (Ts1) and tensile rapture in net section (Ts2).
i)   Tensile yielding, Ts1 = Ag(0.6fy) = 1906 × (0.6 × 250)/1000 = **285** kN
     (IS 800:2007 – 11.2.1)
ii) Tensile rupture, Ts2 = 0.69 Tdn (IS 800:2007 – 11.2.1; 6.3.3)

$$\text{Tdn} = 0.9\,\text{Anc fu} / \gamma_{m1} + \beta\,\text{Agofy} / \gamma_{mo}\left(\text{IS}\,800:2007-6.3.3\right)$$

$$\beta = 1.4 - 0.076\left(w/t\right)\left(fy/fu\right)\left(bs/Lc\right) <= 0.9\,fu\,\gamma_{mo} / fy\,\gamma_{ml} >= 0.7$$

where,
w = outstanding leg width = 100 mm
bs = shear lag width = w (for welded joint) = 100 mm
Lc = length of end connection = 300 mm (IS 800:2007 – 6.3.4)

$$\text{Anc} = \text{net area of the connected leg} = 845\,\text{mm}^2$$

$$\text{Ago} = \text{gross area of outstanding leg} = 1000\,\text{mm}^2$$

t = thickness of the leg = 10 mm
β = lower of the following values of (i) and (ii) but more than or equal to 0.7

$$1.4 - 0.076\left(w/t\right)\left(fy/fu\right)\left(bs/Lc\right)$$

$$= 1.4 - 0.076 \times (100/10) \times (250/410) \times (100/300) = 1.246$$

$$0.9 \text{ fu } \gamma_{mo} / \text{fy} \gamma_{m1} = 0.9 \times 410 \times 1.1 / (250 \times 1.25) = 1.299$$

$$>= 0.7$$

Hence,

$$\beta = \mathbf{1.246}$$

Now,   Tdn $= 0.9 \times 845 \times 410 / 1.25 + 1.246 \times 1000 \times 250 / 1.1$

$$= 532626N = 533 \text{ kN.}$$

Ts2 $= 0.69$ Tdn $= 0.69 \times 533 = \mathbf{368}$ kN

Allowable tension, Ts = 285 kN
d) Compressive Strength of Member, C
Member size: 1SA 100 100 10

Outstanding leg slenderness check : $b/t \le 10.5 \sqrt{(250/\text{Fy})} \left(\text{IS800} - \text{Table 2}\right)$

$$= 10.5 \times \sqrt{(250/250)} = 10.5$$

$$b/t = 100/10 = 10 \text{ Compact section}$$

Allowable compressive strength, $C = \mathbf{0.6\ fcd.Ag} \left(\text{IS800} : 2007 - 11.3.1\right)$

$$= 0.6 \times 158 \times 1903 / 1000$$

$$= \mathbf{180MPa}$$

$$\mathbf{fcd} = \left(\mathbf{fy} / \gamma_{mo}\right) / \left[\phi + \sqrt{\phi^2 - \lambda^2}\right] \left(\text{IS800} : 2007 - 7.1.2.1\right)$$

$$= (250/1.1) / \left[0.92 + \sqrt{(0.92^2 - 0.757^2)}\right] = 158MPa$$

$\lambda =$ slenderness ratio for single angle load through on leg : $\left(\text{IS800} : 2007 - 7.5.1.2\right)$

$$\lambda = \sqrt{\left(\mathbf{k_1} + \mathbf{k_2}\lambda_{vv}^2 + \mathbf{k_3}\lambda_\phi^2\right)}$$

$$= \sqrt{\left(0.2 + 0.35 \times 0.58^2 + 20 \times 0.113^2\right)} = 0.757$$

$$\lambda_{yy} = \left(I / r_{yy}\right) / \left[\varepsilon . \sqrt{\left(\pi^2 E / 250\right)}\right]$$

$$= \left(1000 / 19.4\right) / \left[1 \times \sqrt{\left(3.14^2 \times 200000 / 250\right)}\right] = 0.58$$

$$\lambda_\varphi = \left(b1 + b2\right) / 2t / \left[\varepsilon . \sqrt{\left(\pi^2 E / 250\right)}\right]$$

$$= \left(100 + 100\right) / 2 \times 10 / \left[1 + \sqrt{\left(3.14^2 \times 200000 / 250\right)}\right] = 0.113$$

$$\varepsilon = \sqrt{\left(250 / fy\right)} = 1$$

$$\phi = 0.5 \left[1 + \alpha\left(\lambda - 0.2\right) + \lambda^2\right] = 0.5 \times \left[1 + 0.49 \times \left(0.757 - 0.2\right) + 0.757^2\right] = 0.92$$

where,

	fixed	*hinged*
$k_1=$	0.2	0.7
$k_2=$	0.35	0.6
$k_3=$	20	5

$$l = 1000 \, mm \, \left(assumed\right); r_{vv} = 19.4 mm$$

$$\varepsilon = \sqrt{\left(250 / 250\right)} = 1$$

$$E = 200000 \, MPa$$

$$b_1 = b_2 = 100 mm$$

$$t = 10 \, mm$$

$$\alpha = 0.49$$

Allowable compression, C = 180 kN
e) **Block Shear Capacity, Tb** (IS 800: 2007 – 6.4.1; 6.4.2)
   Member size: **ISA 100 100 10**

$$Lw = 300 mm, \quad t = 10 mm \quad b = 100 mm \quad \left(b_L = b_s = b\right)$$

$$\mathbf{Tdb1} = \left[A_{vg} \, \mathbf{fy} / \left(\sqrt{3} \gamma_{mo}\right) + 0.9 A_{tn} \, \mathbf{fu} / \gamma_{m1}\right]$$

or

$$\mathbf{Tdb2} = \left[\mathbf{0.9A_{vn}\,fu}\,/\left(\sqrt{3}\gamma_{m1}\right) + \mathbf{A_{tg}\,fy}\,/\,\gamma_{mo} \right]$$

Avg = gross area in shear along weld line = $2 \times 300 \times 10 = 6000\text{mm}^2$

Avn = net area in shear along bolt / weld line = 6000mm^2

Atg = minimum gross area in tension perpendicular to the line of force

$$= 100 \times 10 = 1000\text{mm}^2$$

Atn = minimum net area in tension perpendicular to the line of force

$$= 1000\text{mm}^2$$

$$\text{Tdb1} = \left[6000 \times 250 / (1.732 \times 1.1) + 0.9 \times 1000 \times 410 / 1.25 \right] / 1000\text{kN} = 1083\,\text{kN}$$

$$\text{Tdb2} = \left[0.9 \times 6000 \times 410 / (1.732 \times 1.25) + 1000 \times 250 / 1.1 \right] / 1000\text{kN} = 1250\text{kN}$$

$$\text{Tdb} = 1083 \text{ kN; Tb} = 0.69 \text{ Tdb} = 747 \text{ kN}$$

Block shear capacity, Tdb = 747 kN. > Ts
    Let us consider the maximum of tensile and compressive strength for a full strength joint design. (Actual design value of F should be minimum of tension and compression force, when actual length of member is known.)
f) Strength of Fillet Weld Resisting Full Strength (F)
Refer Figure 3.17.

$$\text{Lw} = 300\text{mm} / \text{side}$$

$$\omega = 8\text{mm}$$

$$\text{fyw} = 330\,\text{MPa, faw} = \mathbf{132}\,\text{MPa}$$

Let us consider design strength of weld, fwd = faw/γ_{mw} = 106 MPa
    as per IS 800:2007 *Cl* 10.5.7.1,

$$\text{fwn} = \text{fu} / \sqrt{3} = 540 / 1.732 = 312 \text{ MPa}$$

as per IS 800:2007 *Cl* 11.6.2.4
    The permissible stress of weld, faw = 0.6 fwn = 0.6 × 312 = 187 MPa
    as per IS 800:2007 *Cl* 11.6.3.1

The permissible stress of weld, faw $= 0.4$ fyw $= 0.4 \times 330 = \textbf{132}$ MPa
Strength of fillet weld connecting the single angle to the gusset plate:

$$= 0.7.\omega.\text{fwd} \times 2\text{Lw} \text{ MPa}$$

$$= \textbf{0.7} \times \textbf{8} \times \textbf{106} \times (\textbf{2} \times \textbf{300}) / \textbf{1000} = \textbf{355 kN} > \textbf{F; Safe}$$

3.3.6   DOUBLE ANGLE SECTION – IS 800 (WORKING STRESS)

a) **Material**
   Refer Table 3.1
b) **Design Parameters**
   Refer Figure 3.18
   Member section = **2 ISA 100 100 10**

$$\text{Fillet weld}, \omega = 8 \text{ mm}$$

$$\text{Weld length / side}, \text{Lw} = 250 \text{ mm}$$

$$\text{Gusset thickness}, \text{tg} = 12 \text{ mm}$$

$$\text{Hole dia.for erection bolt}, d = 15.50 \text{ mm}$$

$$4 \text{ nos} (\text{M14})$$

$$b_L = b_s = 100\text{mm} \quad t = 10\text{mm} \quad \text{Ag} = 3806\text{mm}^2 \quad \text{An} = 3651\text{mm}^2$$

c) **Tensile Strength of Member, Ts** (IS 800: 2007 – 11.2)
   Tensile strength, Ts shall be the lower value of tensile yielding in gross section (Ts1) and tensile rapture in net section (Ts2).
   i)   Tensile yielding, Ts1 $=$ Ag $(0.6\text{fy}) = 3806 \times (0.6 \times 250)/1000 = \textbf{571}$ kN
       (IS 800:2007 – 11.2.1)
   ii) Tensile rupture, Ts2 $= 0.69$ Tdn (IS 800:2007 – 11.2.1; 6.3.3)

$$\text{Tdn} = 0.9 \text{Ancfu} / \gamma_{m1} + \beta \text{Agofy} / \gamma_{mo} \ (\text{IS} 800 : 2007 - 6.3.3)$$

$$\beta = 1.4 - 0.076(\text{w} / \text{t})(\text{fy} / \text{fu})(\text{bs} / \text{Lc}) \le 0.9 \text{fu}\gamma_{mo} / \text{fy}\gamma_{m1} \ge 0.7$$

where,
w = outstanding leg width = 100 mm
bs = shear lag width = w (for welded joint) = 100 mm (IS 800:2007 – 6.3.4)
Lc = length of end connection = 250 mm
Anc = net area of the connected leg = 1690 mm²
Ago = gross area of outstanding leg = 2000 mm²

t = thickness of the leg = 10 mm

β = lower of the following values of (i) and (ii) but more than or equal to 0.7

$$1.4 - 0.076 (w/t)(fy/fu)(bs/Lc)$$

$$= 1.4 - 0.076 \times (100/10) \times (250/410) \times (100/250) = 1.215$$

$$0.9 fu\gamma_{mo} / fy\gamma_{m1} = 0.9 \times 410 \times 1.1/(250 \times 1.25) = 1.299$$

(iii) ≥ 0.7

Hence,

$$\beta = 1.215$$

Now,   Tdn $= 0.9 \times 1690 \times 410 / 1.25 + 1.215 \times 2000 \times 250 / 1.1$

$$= 1051160 N = 1051 kN.$$

Ts2 = 0.69 Tdn = $0.69 \times 1051 = 725$ kN

Allowable tension, Ts = 571 kN

d) Compressive Strength of Member, C

Member size: **2 1SA 100 100 10**

Outstanding leg slenderness check : $\mathbf{b/t} <= \mathbf{10.5} \sqrt{(\mathbf{250/Fy})}$ (IS800 – Table 2)

$$= 10.5 \times \sqrt{(250/250)} = 10.5$$

$$b/t = 100/10 = 10 \text{ Compact section}$$

Allowable compressive strength, $\mathbf{C = 0.6fcd.Ag}$ (IS 800 : 2007 – 11.3.1)

$$= 0.6 \times 214 \times 3806 / 1000$$

$$= 489 \, \mathbf{MPa}$$

$$\mathbf{fcd} = (\mathbf{fy}/\gamma_{\mathbf{mo}}) / \left[\phi + \sqrt{(\phi^2 - \lambda^2)} \right] \text{(IS 800 : 2007 – 7.1.2.1)}$$

$$= (250/1.1) / \left[0.58 + \sqrt{(0.58^2 - 0.32^2)} \right] = \mathbf{214} \, \mathrm{MPa}$$

λ = slenderness ratio for double angle struts (IS 800 : 2007 – 7.5.2)

$$\lambda = \sqrt{(\mathbf{f_y}/\mathbf{f_{cc}})}$$

$$= \sqrt{(250/2515)} = \mathbf{0.32}$$

$$\mathbf{f_{cc}} = \pi^2 \mathbf{E} / (\mathbf{KL} / \mathbf{r})^2$$

$$= \left(3.14^2 \times 200000 / 28^2\right) = 2515\,\text{MPa}$$

$$\phi - 0.5\left[1 + \alpha(\lambda - 0.2) + \lambda^2\right] = 0.5 \times \left[1 + 0.49 \times (0.32 - 0.2) + 0.32^2\right] = 0.58$$

where,
L =1000 mm (assumed); $r = r_{xx} = 30.5$ mm
E = 200000 MPa

$$\alpha = 0.49$$

$$\text{KL} / \text{r} = 0.85 \times 1000 / 30.5 = 28$$

Allowable compression, C = 489 kN
e) Block Shear Capacity, Tb (IS 800:2007 – 6.4.1; 6.4.2)
   Member size: **2 ISA 100 100 10**

$$Lw = 250\text{mm}, \quad t = 10\text{mm} \quad b = 100\text{mm} \quad \left(b_L = b_s = b\right)$$

$$\mathbf{Tab1} = \left[\mathbf{A_{vg}}\,\mathbf{fy} / \sqrt{3}\gamma_{\mathbf{mo}}\right) + \mathbf{0.9A_{tn}}\,\mathbf{fu} / \gamma_{\mathbf{m1}}]$$

or

$$\mathbf{Tab2} = \left(\mathbf{0.9A_{vn}}\,\mathbf{fu} / \sqrt{3}\gamma_{\mathbf{m1}}\right) + \mathbf{A_{tg}}\,\mathbf{fy} / \gamma_{\mathbf{mo}})$$

Avg = gross area in shear along weld line = $4 \times 250 \times 10 = 10000\text{mm}^2$

Avn = net area in shear along bolt / weld line = 10000mm^2

Atg = minimum gross area in tension perpendicular to the line of force

$$= 2 \times 100 \times 10 = 2000\text{mm}^2$$

Atn = minimum net area in tension perpendicular to the line of force

$$= 2000\text{mm}^2$$

$$\text{Tdb1} = \left[10000 \times 250 / (1.732 \times 1.1) + 0.9 \times 2000 \times 410 / 1.25\right] / 1000\text{kN} = 1903\text{kN}$$

$$\text{Tdb2} = \left[0.9 \times 10000 \times 410 / (1.732 \times 1.25) + 2000 \times 250 / 1.1\right] / 1000\text{kN} = 2159\text{kN}$$

$$\text{Tdb} = 1903\text{ kN}; \text{Tb} = 0.69\text{ Tdb} = 1313\text{ kN}$$

Block shear capacity, Tdb = 1313 kN. > Ts

Let us consider the maximum of tensile and compressive strength for a full strength joint design. (Actual design value of F, should be minimum of tension and compression force, when actual length of member is known.)

f) Strength of Fillet Weld Resisting Full Strength (F)

Refer Figure 3.17

$$Lw = 250mm / side$$

$$\omega = \mathbf{8}mm$$

$$fyw = 330\,MPa, faw = \mathbf{132}\,MPa$$

Let us consider design strength of weld, $fwd = faw / \gamma_{mw} = \mathbf{106}MPa$

as per IS 800:2007 *Cl* 10.5.7.1,

$$fwn = fu / \sqrt{3} = 540 / 1.732 = 312MPa$$

as per IS 800:2007 *Cl* 11.6.2.4

The permissible stress of weld, $faw = 0.6fwn = 0.6 \times 312 = 187MPa$

as per IS 800:2007 *Cl* 11.6.3.1

The permissible stress of weld, $faw = 0.4fyw = 0.4 \times 330 = \mathbf{132}\,MPa$

Strength of fillet weld connecting the single angle to the gusset plate:

$$= 0.7.\omega.fwd \times \mathbf{4}Lw \qquad Mpa$$

$$= 0.7 \times 8 \times 106 \times (4 \times 250) / 1000 = 597\mathbf{kN} > \mathbf{F}; \mathbf{Safe}$$

3.4 BRACING MEMBER JOINTS

The bracing members may be defined under three categories: Tension, Compression, and Tension-compression members. The joint designer should detail the end connections for full strength (tension/compression) of the member section, unless the member forces are provided in the design drawing. However, the horizontal ties in a bracing system are generally a compression member, and the diagonals in a cross bracing is a tension-compression member.

The rolled steel angles, channels, joists, tubes, and hollow sections are used for bracing member. Connections for angles and joists are discussed in previous chapters. In this section, we will provide end connections with tube and rectangular hollow sections.

3.4.1 Tubular Section

a) Sketch

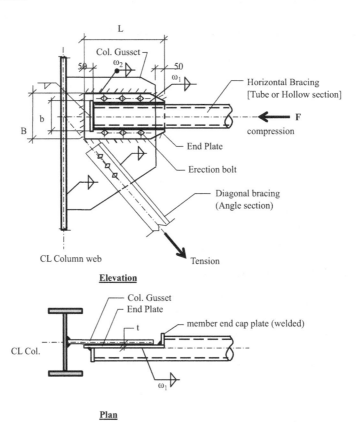

FIGURE 3.19 Bracing end connection.

b) Material

TABLE 3.5
Properties of Tubular Section

IS: 10748

Grade		2	3		
Yield stress	fy	210	240		MPa
Tensile stress	fu	330	410		MPa
Elasticity, E				200000	MPa
Material safety factor	γ_{mo}	1.10	γ_{m1}	1.25	
	γ_{mw}	1.25			
Electrode Ex40-40XX		fuw	540	MPa	
		fyw	330	MPa	

---

c) **Design Parameters**

Member: **Tube 100 M** Nominal weight / m = 12.2 Kg

Outside dia., OD = 114.3 mm No. of erection bolt /side = 3

Pipe thickness, Th = 4.5 mm Hole dia. of erection bolt = 15.5 mm

Area of cross section, A = 1550 mm²; Erection bolt = M14

End Plate: L = 275 mm; B = 255 mm; t = 8 mm

Fillet weld: ω_1 = 6 mm (shop)ω_2 = 6 mm (field)

d) **End Plate**

Area of cross section of member = 1550 mm²

Required cross sectional area of end plate along force, Arqd = 0.5×1550 = 775mm² (assumed 50% for half cut section)

Effective area of cross section of end plate = 255×8 − 3×2×15.5×8

$$= 1296 \text{mm}^2 > \text{Arqd; Safe.}$$

e) **To Determine Allowable Force in the Member, F**

TABLE 3.6
Allowable Comp Strength of Tubular Section

Section	100M	
IS 800:2007 – 7.1.2.1		
r_{min}	38.9	mm
K	0.85	
L	1000	mm
α	0.34	
γ_{mo}	1.10	
fy	210	MPa
E	200000	Mpa
fy/γ_{mo}	191	
λ	0.23	
φ	0.53	
fcd	**189**	MPA
IS 800:2007 – 11.3.1		
fabc	**114**	MPa

Allowable compressive strength, C = fabc A = 114×1550 / 1000 = **177 kN**.

Allowable tensile strength, C = 0.6fy A = 0.6×210×1550 / 1000 = **195kN**.

Design force, F = 177 kN

TABLE 3.7
Buckling Class and Imperfection Factor

Imperfection Factor α

Buck class	a	b	c	d
α	0.21	0.34	0.49	0.76

Buckling Class of Cross-sections

Cross-section	Limits	Bucking About Axis	Buckling Class
Hollow section	Hot-rolled	Any	a
	Cold-formed	Any	b
Channel, angle, T, solids		Any	c

Note: Reference code: IS: 800.

f) Strength of weld

$$Lw = 275 - 50 = 225 \, \text{mm} / \text{side}; L = 275 \, \text{mm}; B = 255 \, \text{mm}$$

$$\omega_1 = 6 \text{mm}; \omega_1 = 6 \text{mm}$$

$$fyw = 330 \, \text{MPa} \, (\text{yield stress}); faw = 132 \, \text{MPa}$$

Let us consider design strength of weld = fwd = **132** MPa.
[*According to IS* 800:2007 *Cl* 10.5.7.1,

$$fwn = fuw / \sqrt{3} = 540 / 1.732 = 312 \text{MPa};$$

According to IS 800:2007 *Cl* 11.6.2.4
The permissible stress of weld, faw = 0.6 fwn = 0.6 × 312 Mpa = 187 MPa;
According to IS 800:2007 *Cl* 11.6.3.1
The permissible stress of weld, faw = 0.4 fyw = 0.4 × 330 = 132 MPa]
Strength of fillet weld, ω_1 (Shop weld)

= Shear strength in weld (along the applied force, F) + tensile resistance in weld

(perpendicular to the axis of force)

$$= 0.7.\omega_1.fwd \times \mathbf{2} \ Lw + 0.7\omega_1.b.(0.5fwd) \quad \text{Mpa}$$

$$= 0.7 \times 6 \times 132 \times (2 \times 225) / 1000 + 0.7 \times 6 \times 134 \times (0.5 \times 132) / 1000$$

$$= \mathbf{287} \text{kN} > \text{F; Safe.}$$

Strength of fillet weld, $\omega2_1$ (Site weld)

= Shear strength in weld $\left(\text{along the applied force}, F\right)$ + tensile resistance in weld

$\left(\text{perpendicular to the axis of force}\right)$

$$= 0.8\left[0.7\omega_2.\text{fwd} \times 2\left(L-50\right) + 0.7\omega_2.b.\left(0.5\text{fwd}\right)\right] \quad \text{Mpa}$$

$$= 0.8\,[0.7 \times 6 \times 132 \times 2\left(275-50\right)/1000 + 0.7 \times 6 \times 255 \times \left(0.5 \times 132\right)/1000$$

$$= \mathbf{256}\,\text{kN} > \text{F; Safe.}$$

3.4.2 RECTANGULAR HOLLOW SECTIONS

a) Sketch
Refer Figure 3.19 above.
b) Material

TABLE 3.8
Properties of Rectangular Hollow Section

IS: 10748				YST310	
Grade		5			
Yield Stress	fy	310			MPa
d or t					
Tensile Stress	fu	490			MPa
Elasticity	E			200000	MPa
Mat. Safety Fac.	γ_{mo}	1.10	γ_{m1}	1.25	
Electrode Ex40-44XX:		fuw	540	Mpa	
	γ_{mw}	1.25	fyw	330	Mpa

---

c) Design Parameters
Member: **Hollow 122 61 3.6**; Nominal weight/m = 9.67 Kg
Depth, D = 122 mm; Wide, B = 61 mm
Thickness, T = 3.6 mm; Area of cross section, A = 1232 mm²
No. of erection bolt /side = 4; Hole dia. of erection bolt = 15.5 mm
Erection bolt = M14

End Plate: L = 400 mm; B = 260 mm; t = 12 mm

$$\text{Fillet weld}: \omega_1 = 6\text{mm}(\text{shop}); \omega_2 = 6\text{mm}(\text{field})$$

d) End Plate

$$\text{Area of cross section of member} = 1232\,\text{mm}^2$$

Required cross sectional area of end plate along force, **Arqd** = 0.5 × 1232 = 616 mm^2
(assumed 50% for half cut section)

$$\text{Effective area of cross section of endplate} = 260 \times 12 - 4 \times 2 \times 15.5 \times 12$$
$$= 1632\,\text{mm}^2 > \text{Areqd; Safe}$$

e) To Determine Allowable Force in the Member, F

TABLE 3.9
Allowable Comp Strength of Rectangular Section

Section	122 61 3.6	
IS 800:2007 – 7.1.2.1		
r_{min}	43.4	mm
K	0.85	
L	1000	mm
α	0.34	
γ_{m0}	1.10	
fy	310	MPa
E	200000	Mpa
fy/γ_{m0}	282	
λ	0.26	
ϕ	0.54	
fcd	**277**	MPa
IS 800: 2007 – 11.3.1		
fabc	**166**	MPa

Allowable compressive strength, C = fabc A = 166 × 1232 / 1000 = 205kN.

Allowable tensile strength, C $= 0.6\,\text{fy A} = 0.6 \times 310 \times 1232 / 1000 = 229\,\text{kN}.$

Design force, $F = 205\,\text{kN}$

f) Strength of Weld

$Lw = 400 - 50 = 350\,\text{mm} / \text{side}; L = 400\,\text{mm}; B = 260\,\text{mm}$

$\omega_1 = 6\,\text{mm}; \omega_2 = 6\,\text{mm}$

$\text{fyw} = 330\,\text{MPa}\,(\text{yield stress}); \text{faw} = 132\,\text{MPa}$

Let us consider design strength of weld = fwd = **132** MPa.
[*According to IS* 800:2007 *Cl* 10.5.7.1,

$$\text{fwn} = \text{fuw} / \sqrt{3} = 540 / 1.732 = 312\,\text{MPa};$$

According to IS 800:2007 *Cl* 11.6.2.4
The permissible stress of weld, faw = 0.6 fwn = 0.6 × 312 MPa = 187 MPa;
According to IS 800:2007 *Cl* 11.6.3.1
The permissible stress of weld, faw = 0.4 fyw = 0.4 × 330 = 132 MPa]
Strength of fillet weld, ω_1 (Shop weld)

= Shear strength in weld (along the applied force, F) + Tensile resistance in weld (perpendicular to the axis of force)

$$= 0.7.\omega_1.\text{fwd} \times 2\,Lw + 0.7\,\omega_1.b.(0.5\,\text{fwd})\,\text{Mpa}$$

$$= 0.7 \times 6 \times 132 \times (2 \times 350) / 1000 + 0.7 \times 6 \times 142 \times (0.5 \times 132) / 1000$$

$$= \textbf{427}\ \text{kN} > F; \text{Safe}.$$

Strength of fillet weld, $\omega2_1$ (Site weld)

= Shear strength in weld (along the applied force, F) + Tensile resistance in weld (perpendicular to the axis of force)

$$= 0.8\left[0.7\omega_2.\text{fwd} \times \textbf{2}(L - 50) + 0.7\omega_2.b.(0.5\,\text{fwd})\right]\text{Mpa}$$

$$= 0.8\left[0.7 \times 6 \times 132 \times 2(400 - 50) / 1000 + 0.7 \times 6 \times 260 \times (0.5 \times 132)\right] / 1000$$

$$= \textbf{368}\,\text{kN} > F; \text{Safe}.$$

3.4.3 Eccentric Joints

Eccentric joints are common in column bracing connections. The gusset connecting bracing members are subject to moment due to eccentricity of member force. An example of design of bracing gusset for eccentric connection is given below for reference.

a) Sketch

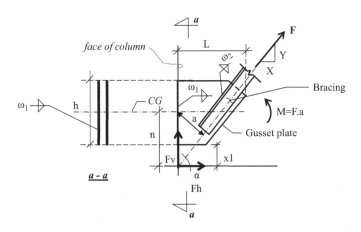

FIGURE 3.20   Detail of eccentric bracing connection.

b) Design Parameters

$$fy = 250\,MPa; Member\,force, F = 100\,kN$$

Bracing chord dimensions:

$$X = 3000\,mm; Y = 3700\,mm$$

Gusset plate dimension:

$$Height, h = 322\,mm; Thickness, t = 12\,mm; x1 = 172\,mm$$

$$Fillet, \omega_1 = 8\,mm; \omega_1 = 8 \times 0.707 = 5.66\,mm\,(effective\,throat\,thickness.)$$

Welding electrode: E70XX

$$Allowable\,welding\,stress, fw = 21\,ksi = 21 \times 6.895 = 145\,MPa$$

$$\alpha = 51\deg\left[\tan\alpha = Y\,/\,X\right]$$

$$n = 333\,mm\left[322 \times 0.5 + 172\right]$$

$$a = 210\text{mm} \left[a = n.\sin(90 - \alpha)^{\circ} \right]$$

$$L = 401\text{mm} \left[L = (h + x1)\ \tan(90 - \alpha)^{\circ} \right]$$

$$Fh = 62.98\text{kN} \left[F\text{Cos}\alpha \right]$$

$$Fv = 77.68\text{kN} \left[F\text{Sin}\alpha \right]$$

$$M = 21\text{kNm} \left[100 \times 210 / 1000 \right]$$

c) Design of Weld, ω_1

$$\text{Shear stress,} Fsh = Fv / (2.h.\omega_1) = 21.31\,\text{N} / \text{mm}^2$$

$$\text{Bending stress,} Fb = M.h. / (2.I_{CG}) = 107.35\,\text{N} / \text{mm}^2$$

$$\text{Tension or compression,} Ft\,\text{or}\,Fc = Fh / (2.h.\omega_1) = 17.28\,\text{N} / \text{mm}^2$$

$$\text{Maxstress} = \sqrt{\left(Fsh^2 + (Fb + Ft)^2 \right)} = 126\,\text{N} / \text{mm}^2 < fw;;\text{Okay.}$$

$$\text{Allowable stress,} fw = 145\,\text{N} / \text{mm}^2$$

d) Design of Gusset

$$\text{Plate size}: h = 322\text{ mm}; t = 12\text{ mm}$$

$$\text{Cross sectional area} = 3864\text{mm}^2 \left[322 \times 12 \right]$$

$$Zxx = 207368\,\text{mm}^2 \left[(12 \times 322^2)/6 \right]$$

$$Ft = Fh / \text{area} = 16\,\text{N} / \text{mm}^2 < 150\text{N} / \text{mm}^2 (\text{allowable stress} 0.6fy)$$

$$Fb = M / Zxx = 101\text{N} / \text{mm}^2 < 165\text{N} / \text{mm}^2 (\text{allowable stress} 0.66fy)$$

$$Fsh = \frac{Fv}{\text{area}} = 20\text{N} / \text{mm}^2 < 100\text{N} / \text{mm}^2 (\text{allowable stress} 0.4fy)$$

$$Fc = \frac{Fh}{\text{area}} = 16\text{N} / \text{mm}^2 < 51\text{N} / \text{mm}^2 (\text{allowable stress} Pc)$$

Allowable compressive stress:

$$Kl = 0.7L = 280mm$$

$$r_{min} = 3.6mm$$

$$Kl / r_{min} = 78; \quad Pc = 51N / mm^2 \left(IS\,800 : 2007 - 7.1.2.1\right)$$

$$Fc / Pc + Fb / Pb = \left(16 / 51\right) + \left(101 / 165\right) = 0.93 < 1; Okay.$$

Method of design of member end connecting weld, fillet, ω_2 is shown in Section 3.3 Truss member Joint.

3.5 SPLICE JOINT – PLATED AND ROLLED SECTION

The welded splice for beams and columns is generally required to extend the length of member as per drawing requirement. It is preferable that shop splices should be welded joint and field splices are bolted.

The workout examples given in this section include splice of beam, plate girder, and channel section in accordance with AISC and IS code. These examples are also applicable for column shafts, unless those are full penetration but welded joints as shown in Chapter 6 of this book.

3.5.1 BEAM AND GIRDER SPLICE – AISC (ASD)

a) **Sketch**
   Refer Figure 3.21.
b) **Material**
   See Table 3.10.
c) **Member Descriptions**

Spliced member:	W24 × 84	Type – 1
Flange:	Width, bf = 9.02 inch	Thickness, tf = 0.77 inch
Web:	Depth, d = 24.1 inch	Thickness, tw = 0.47 inch

d) **Design Parameters**
   (i) Flange force, F: Full tension capacity = 0.6 fy. Af, where, Af = flange area
   (ii) Web shear, V: 70% of web shear capacity, where, Aw = web area
   (iii) Sum area of cover plates is 5% more than the area of the section spliced.
   (iv) All welds are shop welds.

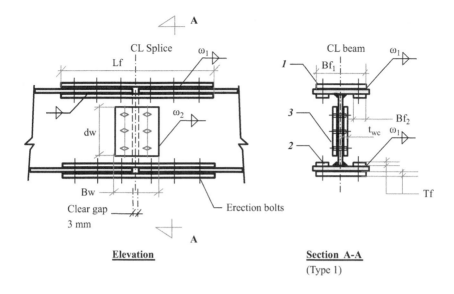

FIGURE 3.21 Plate girder splice – welded.

TABLE 3.10
(Refer: AISC Manual)

ASTM A 36

Yield stress, fy	F_{nBM}	36	ksi	250	MPa
Tensile stress	F_{uBM}	65	ksi	450	MPa
Elasticity	Es			200000	MPa
Electrode Fe 70		70	ksi		
Nominal stress	F_{nw}	42	ksi	290	MPa
(Nominal stress = 0.6 Fexx)					

e) Sizing of Flange Cover Plate and Weld ω_1:

Sizing of flange cover plate and weld:

Flange cover plate – 1:	Flange cover plate – 2:
Wide, $Bf_1 = 8.00$ inch	Wide, $Bf_2 = 2.77$ inch
Thick, $Tf = 0.63$ inch	Thick, $Tf = 0.63$ inch
Length, $Lf = 24.00$ inch	Length, $Lf = 24.00$ inch
Weld, $\omega_1 = 0.5$ inch	Wwld, $\omega_1 = 0.25$ inch
Nos per flange = 1	Nos per flange = 2

Cross sectional area of flange, $Af = 9.02 \times 0.77 = 6.95 \, \text{inch}^2$

Required sectional area of cover plate, $A \, \text{required} = Af \times 1.05 = 7.3 \, \text{inch}^2$

Total sectional area of cover plate, $A \, \text{provide} = 1 \times 8 \times 0.625 + 2 \times 2.77 \times 0.625$
$$= 8.46 \, \text{inch}^2 > A \, \text{required}; \text{Okay.}$$

Welding to cover plate 1 (one per flange)
Max axial load in plate 1,

$$F1 = 0.6 \, fy.Bf_1.Tf / \Omega$$

$$= 0.6 \times 36 \times 8 \times 0.625 / 1.5$$

$$= 72 \, \text{kips.}$$

$$\left[\Omega = 1.5 \, \text{for base material} \right]$$

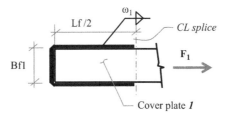

FIGURE 3.22   Plan view of fillet weld.

Strength of fillet weld connecting the cover plate 1 to the flange of spliced beam:

$$= \left[0.7.\omega_1.Fnw \times 2 (Lf / 2) + 0.7.\omega_1.(0.5 Fnw).Bf1 \right] / \Omega \, \text{kips}$$

$$= \left[0.7 \times 0.5 \times 42 \times 24 + 0.7 \times 0.5 \times 0.5 \times 42 \times 8 \right] / 2$$

$$= 206 \text{ kips} \quad > \text{F1; Safe.}$$

[50% strength considered for weld perpendicular to the axis of force.]

$$\Omega = 2 \quad \left(\text{AISC table J2.5} \right)$$

Welding to cover plate 2 – (2 per flange)
Max axial load in each plate,

$$F2 = 0.6 \text{fy}.\text{Bf2}.\text{Tf} / \Omega$$

$$= 0.6 \times 36 \times 2.77 \times 0.625 / 15$$

$$= 24.67 \text{kips.}$$

$$\Omega = 1.5 \quad (\text{AISC table J2.5})$$

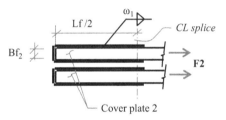

FIGURE 3.23   Top plan.

Strength of fillet weld connecting the cover plate 2 to the flange of spliced beam:

$$= \left[0.7.\omega_1.\text{Fnw} \times 2 \left(\text{Lf} / 2 \right) + 0.7.\omega_1.\left(0.5\text{Fnw} \right).\text{Bf2} \right] / \Omega \text{ kips}$$

$$= \left[0.7 \times 0.25 \times 42 \times 24 + 0.7 \times 0.25 \times 0.5 \times 42 \times 2.77 \right] / 2$$

$$= 93 \text{kips} \quad > \text{F}_1; \text{Safe.}$$

$$\Omega = 2 \left(\text{AISC table J2.5} \right)$$

f) Sizing of Web Cover Plate and Weld ω_2
Web cover/splice plate – 3 (on both sides)

$$\text{Wide}, \text{Bw} = 10.00 \text{ inch}$$
$$\text{Depth}, \text{dw} = 19.00 \text{ inch}$$

$$\text{Thickness}, t_{wc} = 0.5 \, \text{inch}$$

$$\text{Weld}, \omega_2 = 0.25 \, \text{inch}$$

$$\text{Web area of spliced beam}, Aw = 11.33 \, \text{inch}^2$$

$$\left(24.1 \times 0.47\right) \quad \frac{\text{add } 5\% = 0.57 \, \text{inch}^2}{\text{Sum} = 11.90 \, \text{inch}^2}$$

Provide two splice plates resisting web shear, Asp

$$\text{Asp} = 2 \times 19 \times 0.5 = 19 \text{inch}^2 > Aw; \text{Safe}$$

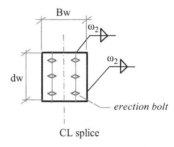

FIGURE 3.24   Web cover plate (one per side).

Welding of cover plate 3 – (on both sides), ω_2

$$\text{Web shear}, V = 70\% \left(0.6 \text{fy} \, Aw\right) / \Omega$$

$$= 0.7 \times 0.6 \times 36 \times 24.1 \times 0.47 / 1.6$$

$$= 103 \, \text{kips}$$

$$\Omega = 1.67$$

$$\mathbf{R = V / 2 = 51.5 \, kips}$$

In Figure 3.25,

$$n = b^2 / \left(2b + Lv\right)$$

$$Ch = Lv - n - \left(1 / 16\right)$$

$$Cv = Lv / 2$$

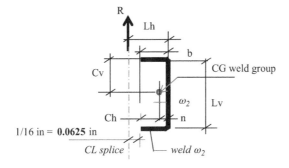

FIGURE 3.25 Welding of cover plate 3.

$$b = Lh - (1/16)$$

$$Jw = \left[(2b + Lv)^3 / 12 - b^2 (b + Lv)^2 / (2b + Lv) \right]$$

Twisting (horizontal)

$$fh = TCv / Jw = \left[R(Lh - n)Cv / 2.Jw \right]$$

Twisting (vertical)

$$fv1 = TCh / Jw = \left[R(Lh - n)Ch / 2.Jw \right]$$

Shear (vertical)

$$fv2 = \left[0.5R / (2b + Lv) \right] \quad fr = \sqrt{\left[fh^2 + (fv1 + fv2)^2 \right]}$$

In this example,

$$b = 4.94 \, inch \, (5 - 0.0625); R = 52 \, kips \quad Lv = 19 \, inch \quad Lh = 5 \, inch$$

$$n = 0.85 \, inch \left[494^2 / (2 \times 4.94 + 19) \right]$$

$$Ch = 4.09 \, inch \, (5 - 0.85 - 0.0625) \quad Cv = 9.5 \, inch \, (19/2)$$

$$Jw = 2007 - 484 = 1523 \, inch^3$$

$$fh = 0.667 \, kips / inch \left[51.5 \times (5 - 0.85) \times 9.5 / (2 \times 1523) \right]$$

$$fv1 = 0.287 \, kips / inch \left[51.5 \times (5 - 0.85) \times 4.09 / (2 \times 1523) \right]$$

$$fv2 = 0.892 \text{kips / inch} \left[0.5 \times 51.5 / \left(2 \times 4.94 + 19 \right) \right]$$

$$\mathbf{fr} = \sqrt{\left[0.667^2 + \left(0.287 + 0.892 \right)^2 \right]} = \mathbf{1.36} \, \text{kips / inch}.$$

Provide fillet weld of size, $\omega_2 = 0.25$ inch

Strength of fillet weld connecting cleat angles to the web of supported beam at shop

$$= 0.7.\omega2.\text{Fnw} / \Omega \, \text{kips / inch} = 0.7 \times 0.25 \times 42 / 2 = 3.68 \text{kips / inch}$$

$$> \text{fr; Safe.}$$

$$\Omega = 2(\text{AISC Table J2.5})$$

3.5.2   SPLICE FOR CHANNEL SECTION – AISC (ASD)

a) **Sketch**

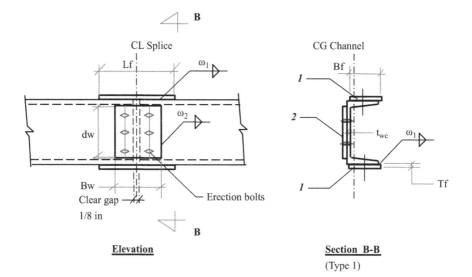

Elevation   **Section B-B**
                                                                                 (Type 1)

FIGURE 3.26   Channel member splice – welded.

b) **Material**
   Refer Table 3.11.
c) **Member Descriptions**
   Spliced member: **C 10 × 15.3**

Flange –	wide, bf = 2.6 inch	thickness, tf = 0.44 inch
Web –	depth, d = 10 inch	thickness, tw = 0.24 inch

TABLE 3.11
Column Base Materials (AISC/ACI)

Structural Steel: ASTM A36

Yield stress	Fy	36	ksi	250	MPa
Tensile stress	Fu	65	ksi	450	MPa
Elasticity	Es	29000	ksi	200000	MPa
Anchor bolt: A36 (ACI 318-19)					
	fy	36	ksi	250	MPa
Concrete Pedestal: (ACI 318-19)					
Conc. strength	fc'	3600	psi	25	MPa
	Ec	3420000	psi	23586	MPa
	m			11	

d) Design Parameters

i) Flange force, F: Full tension capacity = 0.6 fy. Af, where, Af = flange area

ii) Web shear, V: 70% of web shear capacity, where, Aw = web area

iii) Sum area of cover plates is 5% more than the area of the section spliced.

iv) All welds are shop weld.

e) Sizing of Flange Cover Plate and Weld ω_1:

Sizing of flange cover plate and weld:

Flange cover plate – 1:

$$\text{Wide}, Bf_1 = 3.00\,\text{inch}$$

$$\text{Thick}, Tf = 0.50\,\text{inch}\,(1/2\,\text{inch})\,\text{Length}, Lf = 10.00\,\text{inch}$$

$$\text{Weld}, \omega_1 = 0.25\,\text{inch}\,(1/4\,\text{inch}); \text{Nos per flange} = 1$$

$$\text{Cross sectional area of flange}, Af = 2.6 \times 0.44 = 1.14\,\text{inch}^2$$

$$\text{Required sectional area of cover plate}, A\,\text{required} = Af \times 1.05 = 1.2\,\text{inch}^2$$

$$\text{Total sectional area of cover plate}, A\,\text{provided} = 1 \times 3 \times 0.5 = 1.5\,\text{inch}^2$$

$$> A\,\text{required}; \text{Okay}.$$

Welding to cover plate 1 *(one per flange)*
See Figure 3.22, Plan view of fillet weld

Max axial load in plate 1,

$$F_1 = 0.6\,fy.Bf_1.Tf / \Omega$$

$$= 0.6 \times 36 \times 3 \times 0.5 / 1.5$$

$$= 21.33\,kips. \Omega = 1.5$$

Strength of fillet weld connecting the cover plate 1 to the flange of spliced beam:

$$= \left[0.7.\omega_1.Fnw \times 2\left(Lf / 2\right) + 0.7.\omega_1.\left(0.5Fnw\right).Bf1\right] / \Omega\,kips$$

$$= \left[0.7 \times 0.25 \times 42 \times 10 + 0.7 \times 0.25 \times 0.5 \times 42 \times 3\right] / 2$$

$$= 42\,kips > F1; Safe.$$

[50 % strength considered for weld perpendicular to the axis of force.]

$$\Omega = 2\left(AISC\ Table\ J2.5\right)$$

f) Sizing of Web Cover Plate and Weld ω_2

Web cover/splice plate

Wide, $Bw = 6.00$ inch; depth, $dw = 8.00$ inch thickness, $t_{wc} = 0.50$ inch $\left(1/2 inch\right)$

$$Weld, \omega_2 = 0.25\,inch\,\left(1/4 inch\right)$$

$$Web\ area\ of\ spliced\ beam, Aw = 2.4\,inch^2$$

$$\left(10 \times 0.24\right) \quad \underline{add\ 5\% = 0.12\,inch^2}$$
$$Sum = 2.52\,inch^2$$

Provide splice plate of area, $Asp = 8 \times 0.5 = 4.00\,inch^2 > Aw, Safe.$

Welding of cover plate 2 – ω_2
Refer Figure 3.25 above

Web shear, $V = 70\% \left(0.6\,fy\,Aw\right) / \Omega = 0.7 \times 0.6 \times 36 \times 10 \times 0.24 / 1.67 = \mathbf{22}\,kips$

$$\Omega = 1.67$$

$$\mathbf{R = V / 2 = 11\,kips}$$

In Figure 3.25,

$$n = b^2 / (2b + Lv)$$

$$Ch = Lv - n - (1/16)$$

$$Cv = Lv / 2$$

$$b = Lh - (1/16)$$

$$Jw = \left[(2b + Lv)^3 / 12 - b^2 (b + Lv)^2 / (2b + Lv) \right]$$

Twisting (horizontal)

$$fh = TCv / Jw = \left[R(Lh - n)Cv / 2.Jw \right]$$

Twisting (vertical)

$$fv1 = TCh / Jw = \left[R(Lh - n)Ch / 2.Jw \right]$$

Shear (vertical)

$$fv2 = \left[0.5R / (2b + Lv) \right]; fr = \sqrt{ \left[fh^2 + (fv1 + fv2)^2 \right] }$$

In this example,

$$b = 2.94 \text{ inch} (3 - 0.0625); R = 11 \text{ kips} \quad Lv = 8 \text{ inch} \quad Lh = 3 \text{ inch}$$

$$n = 0.62 \text{ inch} \left[2.94^2 / (2 \times 2.94 + 8) \right]$$

$$Ch = 2.32 \text{ inch} (3 - 0.62 - 0.0625)Cv = 4 \text{ inch} (8/2)$$

$$Jw = 223 - 75 = 148 \text{ inch}^3$$

$$fh = 0.349 \text{ kips / inch} \left[11 \times (3 - 0.62) \times 4 / (2 \times 148) \right]$$

$$fv1 = 0.203 \text{ kips / inch} \left[11 \times (3 - 0.62) \times 2.32 / (2 \times 148) \right]$$

$$fv2 = 0.391 \text{ kips / inch} \left[0.5 \times 11 / (2 \times 2.94 + 8) \right]$$

$$\mathbf{fr} = \sqrt{\left[0.349^2 + \left(0.203 + 0.391\right)^2\right]} = \mathbf{0.69}\,\text{kips / inch.}$$

Provide fillet weld of size, $\omega_2 = 0.25$ inch

Strength of fillet weld connecting cleat angles to the web of supported beam at shop

$$= 0.7\omega2.\text{Fnw} / \Omega\,\text{kips / inch} = 0.7 \times 0.25 \times 42 / 2 = 3.68\,\text{kips / inch} > \text{fr; Safe.}$$

$$\Omega = 2\left(\text{AISC Table J2.5}\right)$$

3.5.3 BEAM AND GIRDER SPLICE – IS 800 (WORKING STRESS)

a) **Sketch**
   Refer Figure 3.21
b) **Material**
   Same as in Table 3.1 above
c) **Member Descriptions**
   Spliced member: **ISMB 600**

Flange –	wide, bf = 210 mm	thickness, tf = 20.8 mm
Web –	depth, d = 600 mm	thickness, tw = 12 mm

d) **Design Parameters**
   i)   Flange force, F: Full tension capacity = 0.6 fy. Af; where, Af = flange area.
   ii) Web shear, V: 70% of web shear capacity (= 0.4 fy Aw); where, Aw = web area.
   iii) Sum area of cover plates is 5% more than the area of member section spliced.
   iv) All welds are shop weld.
e) **Sizing of Flange Cover Plate and Weld**:
   Let us assume following dimensions for flange cover plates,

Flange cover plate – 1:	Flange cover plate – 1:
Wide, $Bf_1 = 185$ mm	Wide, $Bf_2 = 65$ mm
Thick, $Tf_1 = 25$ mm	Thick, $Tf_2 = 12$ mm
Length, Lf = 660 mm	Length, Lf = 660 mm
Weld, $\omega_1 = 12$ mm	Weld, $\omega_1 = 12$ mm
Nos per flange = **1**	Nos per flange = **2**

Cross sectional area of flange, $Af = 210 \times 20.8 = 4368\,\text{mm}^2$

Required sectional area of cover plate, A required $= Af \times 1.05 = 4586\,\text{mm}^2$

Total sectional area of cover plate, A provided $= 1 \times 185 \times 25 + 2 \times 65 \times 12 = 6185 \text{mm}^2$

$$> \text{A required; Okay.}$$

Tension in cover plate:
   The actual tensile stress on the cover plate shall be smaller of the following:

a) Yielding of gross section, fat = 0.6 fy (IS 800 : 2007 11.2.1)
b) Rupture of the net section, fat = 0.69 Tdn/Ag
c) Block shear resistance, fat = 0.69 Tdb/Ag

a) *By yielding*:

$$\text{fat} = 0.6 \times 250 = \mathbf{150} \text{ MPa}$$

b) *By rapture*:

$$\text{Tdn} = 0.9 \text{Anfu} / \gamma_{m1} = 0.9 \times 4625 \times 410 / 1.25 / 1000$$
$$= 1365 \text{kN} \left(\text{IS } 800 : 2007\,6.3.1 \right)$$

$$\text{An} = 185 \times 25 = 4625 \text{mm}^2 = \text{Ag} \left(\text{without bolt holes} \right)$$

$$\text{fat} = 0.69 \times 1365000 / 4625 = 204 \text{MPa}$$

c) *By block shear*:

Cover plate $1 - \text{Lw} = 330 \text{ mm}; \text{b} = 185 \text{ mm}; \text{t} = 25 \text{ mm}$
$$\text{Tdb} = \text{Smaller of Tdb1 or Tdb2} \left(\text{IS } 814 : 2004 - 6.4.1; 6.4.2 \right)$$

$$\mathbf{Tdb1} = \left[\mathbf{A_{vg}\, fy} / \left(\sqrt{3} \gamma_{mo} \right) + \mathbf{0.9 A_{tn}\, fu} / \gamma_{m1} \right]$$

or

$$\mathbf{Tdb2} = \left[\mathbf{0.9 A_{vn}\, fu} / \left(\sqrt{3} \gamma_{m1} \right) + \mathbf{A_{tg}\, fy} / \gamma_{mo} \right]$$

Avg = gross area in shear along weld line $= 2 \times 330 \times 25 = 16500 \text{mm}^2$

Avn = net area in shear along weld line $= 16500 \text{mm}^2$

Atg = minimum gross area in tension perpendicular to the line of force
$$= 185 \times 25 = 4625 \text{mm}^2$$

Atn = minimum net area in tension perpendicular to the line of force = 4625mm^2

$$Tdb1 = \left[16500 \times 250 / (1.732 \times 1.1) + 0.9 \times 4625 \times 410 / 1.25\right] / 1000 = 3530 kN$$

$$Tdb2 = \left[0.9 \times 16500 \times 410 / (1.732 \times 1.25) + 4625 \times 250 / 1.1\right] / 1000 = 3863 kN$$

$$Tdb = 3530 \, kN; fat = 0.69 \, Tdb \, / \, Ag = \mathbf{527} \, kN$$

$$f_{at} \text{ for design} = \text{minimum of } f_{at} \text{ in a), b) and c) above} = \mathbf{150} \, MPa$$

$$\text{Axial force in cover plate 1, F1} = 150 \times 185 \times 25 / 1000 = \mathbf{694} kN$$

Welding to cover plate 1 (one per flange)
Refer Figure 3.22 Plan view of fillet weld

$$F1 = \mathbf{694} \, kN; Lf = 660 \, mm; Bf1 = 185 \, mm$$

$\omega_1 = 12 \, mm$; fyw = 330 MPa (yield stress); faw = 132 MPa
Let us consider design strength of weld = fwd = faw = **132** MPa
as per IS 800:2007 *Cl* 10.5.7.1,

$$fwn = fu / \sqrt{3} = 410 / 1.732 = 237 \, MPa$$

as per IS 800:2007 *Cl* 11.6.2.4
The permissible stress of weld, faw = 0.6 fwn = 0.6 × 237 = 142 MPa
as per IS 800:2007 *Cl* 11.6.3.1
The permissible stress of weld, faw = 0.4 fyw = 0.4 × 330 = **132** MPa
Strength of fillet weld connecting the cover plate 1 to the flange of spliced beam:

$$= \left[0.7.\omega_1.fwd \times 2(Lf / 2) + 0.7.\omega_1.(0.5fwd).Bf1\right] / 1000 \, kN$$

$$= \left[0.7 \times 12 \times 132 \times 660 + 0.7 \times 12 \times 0.5 \times 132 \times 185\right] / 1000$$

$$= \mathbf{834} kN > F1; Safe.$$

[50% strength considered for weld perpendicular to the axis of force.]
Welding to cover plate 2 (2 per flange)
Refer Figure 3.23.

max axial load in each plate, F2 = fat.Bf$_2$.Tf = 150 × 65 × 12 / 1000 = 117 kN

Strength of fillet weld connecting the cover plate 2, to the flange of spliced beam:

$$= \left[0.7.w_1.fwd \times 2\left(Lf/2\right)+0.7.w_1.\left(0.5fwd\right).Bf_2\right]/1000kN$$

$$= \left[0.7 \times 12 \times 132 \times 660 + 0.7 \times 12 \times 0.5 \times 132 \times 65\right]/1000$$

$$= \mathbf{768}kN > Fl; Safe.$$

[50% strength considered for weld perpendicular to the axis of force.]
Refer Figure 3.24
Web cover/splice plate – 3 (on both sides)

$$wide, Bw = 240mm; depth, dw = 480\,mm; thickness, t_{wc} = 10mm$$

$$\omega_2 = 8mm$$

$$Web\,area\,of\,spliced\,beam, Aw = \left(600 \times 12\right) = 7200mm^2$$

$$Add\ 5\% = 360\ mm^2$$
$$Total\ =\ 7560\ mm^2$$

Provide two splice plates resisting web shear, $Asp = 2 \times 480 \times 10$
$$= 9600mm^2 > Aw; Safe.$$

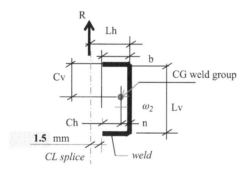

FIGURE 3.27 Welding of cover plate 3.

$$Web\,shear, V = 70\%\left(0.4fy\,Aw\right)/1000 = 0.7 \times 0.4 \times 250 \times 600 \times 12 = 504kN$$

$$R = V/2 = \mathbf{252}kN$$

In Figure 3.27,

$$n = b2/\left(2b+Lv\right)$$

$$Ch = Lv - n - 1.5$$

$$CvLv / 2$$

$$b = Lh - 1.5$$

$$Jw = \left[(2b + Lv)^3 / 12 - b_2 (b + Lv)^2 / (2b + Lv) \right]$$

Twisting (horizontal)

$$fh = TCv / Jw = \left[R(Lh - n)Cv / 2.Jw \right]$$

Twisting (vertical)

$$fv1 = TCh / Jw = \left[R(Lh - n)Ch / 2.Jw \right]$$

Shear (vertical)

$$fv2 = \left[0.5R / (2b + Lv) \right]; \quad \mathbf{fr} = \sqrt{\left[\mathbf{fh2} + (\mathbf{fv1} + \mathbf{fv2})^2 \right]}$$

In this example,

$$b = 118.5 \text{ mm} \sim (120 - 1.5); R = 252000 \text{ N}$$

$$Lv = 480 \text{ mm}; Lh = 120 \text{ mm}$$

$$n = 19.58 \text{mm} \quad \left[118.5^2 / (2 \times 118.5 + 480) \right]$$

$$Ch = 98.92 \text{ mm} (120 - 19.58 - 1.5)$$

$$Cv = 240 \text{ mm} (480 / 2)$$

$$Jw = 30716818 - 7015294 = 23701524 \text{mm}^3$$

$$fh = 128 \text{N} / \text{mm} \left[252000 \times (120 - 19.58) \times 240 / (2 \times 23701524) \right]$$

$$fv1 = 53 \text{N} / \text{mm} \left[252000 \times (120 - 19.58) \times 98.92 / (2 \times 23701524) \right]$$

$$fv2 = 176 \text{N} / \text{mm} \left[0.5 \times 252000 / (2 \times 119 + 480) \right]$$

$$fr = \sqrt{\left[128^2 + (53 + 176)^2 \right]} = \mathbf{262} \text{N} / \text{mm}$$

Provide fillet weld of size, $\omega_2 = 8$ mm

Strength of fillet weld connecting cleat angles to the web of supported beam at shop:

$$= 0.7.\omega2.\text{fwd}\,N / mm$$

$$= 0.7 \times 8 \times 132 = 739N / mm \qquad > fr \,; \text{Safe.}$$

3.5.4 SPLICE FOR CHANNEL SECTION – IS 800 (WORKING STRESS)

a) Sketch

Refer to Figure 3.26

b) Materials

See Table 3.1.

c) Member Descriptions

Splice member:	**ISMC 250**	Type – 1
Flange:	Width, bf = 80 mm	Thickness, tf = 14.1 mm
Web:	Depth, d = 250 mm	Thickness, tw = 7.1 mm

d) Design Parameters

i)   Flange force, F: Full tension capacity = 0.6 fy. Af, where, Af = flange area/

ii) Web shear, V: 70% of web shear capacity, where, Aw = web area.

iii) Sum area of cover plates is 5% more than the area of the section spliced.

iv) All welds are shop weld.

e) Sizing of Flange Cover Plate and Weld ω_1:

Let us assume the flange cover plate and weld of following sizes:

Flange cover plate – 1:

Wide, Bf_1 = 90 mm	Thick, Tf = 16 mm
Length, Lf = 400 mm	Weld, ω_1 = 8 mm

Nos per flange = 1

$$\text{Cross sectional area of flange}, Af = 80 \times 14.1 = 1128 \text{mm}^2$$

$$\text{Required sectional area of cover plate}, A \text{ required} = Af \times 1.05 = 1184 \text{mm}^2$$

$$\text{Total sectional area of cover plate}, A \text{ provided} = 1 \times 90 \times 16 = 1440 \text{mm}^2$$

$$> A \text{ required}; \text{Okay.}$$

Welding to cover plate 1 (one per flange) – ω_1,
   Refer Figure 3.21 above
   Max axial load in plate 1,

$$F1 = 0.6fy.A \text{ provided} = 0.6 \times 250 \times 1440 / 1000 = 216kN$$

$$Lf = 400mm \qquad Bf = 90mm \qquad \omega_1 = 8mm$$

$$fyw = faw = \mathbf{132}MPa$$

[as per IS 800:2007 Cl 10.5.7.1, fwn = fu/ $\sqrt{3}$ = 410/1.723 = 237 MPa
   and as per IS 800:2007 Cl 11.6.2.4,

   The permissible stress of weld, faw = 0.6fwn = 0.6×237 = 142MPa

as per IS 800:2007 Cl 11.6.3.1

   The permissible stress of weld, faw = 0.4fyw = 0.4×330 = **132**MPa]

Strength of fillet weld connecting the cover plate 1 to the flange of spliced beam:

$$= \left[0.7.\omega_1.fwd \times 2\left(Lf/2\right) + 0.7.\omega_1.\left(0.5fwd\right).Bf1\right]/1000 \text{ kN}$$

$$= \left[0.7 \times 8 \times 132 \times 400 + 0.7 \times 8 \times 0.5 \times 132 \times 90\right]/1000$$

$$= 329kN \qquad > F1; Safe.$$

[50% strength considered for weld perpendicular to the axis of force.]
f) Sizing of Web Cover Plate and Weld, ω_2.
   Web cover/splice plate

Wide, Bw = 100 mm	Depth, dw = 200 mm
Thickness, t_{wc} = 12 mm	Weld, ω_2 = 8 mm

Web area of spliced beam, Aw = 250×7.1 = 1775mm^2

$$\text{Add } 5\% = 89mm^2$$

$$\overline{\text{Total Aw} = 1864mm^2}$$

Provide splice plate of area, $Asp = 200 \times 12 = 2400 mm^2 < Aw; Safe.$

Welding of cover plate 2, ω^2

Web shear, $V = 70\% (0.4 fy\, Aw) / 1000 = 0.7 \times 0.4 \times 250 \times 250 \times 7.1 = 124 kN$

$$R = V / 2 = \mathbf{62} kN$$

Refer Figure 3.27 above.
   In this example,

$$b = 48.5 \, mm \, (50 - 1.5); R = 62000 \, N$$

$$Lv = 200 \, mm; Lh = 50 \, mm$$

$$n = 7.92 \, mm \left[48.5^2 / (2 \times 48.5 + 200) \right]$$

$$Ch = 40.58 \, mm \, (50 - 7.92 - 1.5)$$

$$Cv = 100 \, mm \, (200 / 2)$$

$$Jw = 2183173 - 489080 = 1694093 mm^3$$

$$fh = 77 N / mm \quad \left[62125 \times (50 - 7.92) \times 100 / (2 \times 1694093) \right]$$

$$fv1 = 31 N / mm \quad \left[62125 \times (50 - 7.92) \times 40.58 / (2 \times 1694093) \right]$$

$$fv2 = 105 N / mm \quad \left[0.5 \times 62125 / (2 \times 48.5 + 200) \right]$$

$$fr = \sqrt{ \left[77^2 + (31 + 105)^2 \right] } = \mathbf{156} N / mm$$

Provide fillet weld of size, $\omega_2 = 8 \, mm$

Strength of fillet weld connecting cleat angles to the web of supported beam at shop:

$$= 0.7 . \omega2 . fwd \, N / mm$$

$$= 0.7 \times 8 \times 132 = \mathbf{739} N / mm \quad > fr; Safe.$$

3.6 COLUMN BASES

Column bases are a connection between the steel column section and the reinforced concrete pedestal. There are workout examples of various types of column bases in my book, "Design of Industrial Structures – RCC and Steel." In this section, workout examples of a single leg built-up column and a two-legged column are provided for the reader's reference.

3.6.1 SINGLE LEG BUILT-UP I-SECTION – AISC

a) **Materials**
   Refer Table 3.11.
b) **Design Parameters**
   Column size:

$$Dc = 450 \, mm$$

$$Wc = 150 \, mm$$

   Pedestal:

$$Dp = 950 \, mm$$

$$Wp = 600 \, mm$$

   Base Plate:

$$Width, B = 400 \, mm$$

$$Length, N = 750 \, mm$$

$$Thickness, t = 25 \, mm$$

   Anchor bolt:

$$Nominal \, dia, \, \phi_{dia} = 24mm$$

$$Nos / side, n = 3$$

$$Dist. from \, col.face = 100 \, mm$$

$$Edge \, distance, x = 50 \, mm$$

$$Spacing, s = 133 \, mm$$

Stiffener plate, a:

$$\text{Height} = 350\,\text{mm}$$

$$\text{Thickness} = 12\,\text{mm}$$

$$\text{Nos} = 2$$

$$\omega = 8\,\text{mm}\,(\text{fillet weld})$$

Stiffener plate, b:

$$\text{Height} = 250\,\text{mm}$$

$$\text{Thickness} = 12\,\text{mm}$$

$$\text{Nos} = 4$$

Shear lug:
Depth, L = 125 mm; wide, b = 150 mm; thick, t_1 = 25 mm
Nos = 1 Grout thickness = 40 mm
Load resistant factors:
For moment and axial load: ASD $\Omega = 1.67$ LRFD $\phi = 0.9$
For bearing: ASD $\Omega = 1.67$ LRFD $\phi = 0.65$
Design Load: *(Unfactored)*
Axial comp, P = 300 kN; Moment about major axis, Mx = 150 kNm
Shear along Y-axis, Fx = 30 kN
c) Strength Design
Eccentricity check

$$e = M / P = 150 / 300 = 0.5\,\text{m} = 500\,\text{mm}$$

$$\text{Length of base plate,} N = 750\,\text{mm}\quad L/6 = 125\,\text{mm}$$

$$e > L/6 = 125\,\text{mm; tension occurs at base.}$$

If, e (=M/P) is less than L/6, i.e., resultant vertical load falls within middle third, there will be no tension in base plate/anchor bolt. Otherwise, tension occurs between bottom of base plate and top of concrete pedestal.

At this plane, the force transfer from steel column base and pedestal will be similar to a reinforced concrete beam with reinforcement (here anchor bolt) carrying tensile force and at the opposite end compression load transferred by surface contact between base plate grout.

N A depth, Y:
Y = m. fc. d / (m fc + fs)

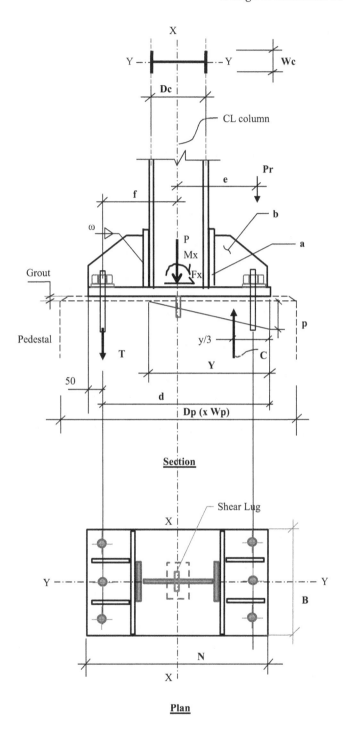

FIGURE 3.28 Column base – Single column.

Let us consider,

$$m = 11$$

$$fc = 0.85fc' = 0.85 \times 25 = 21.25 MPa$$

$$fs = 0.6fy = 0.6 \times 250 = 150 MPa$$

$$d = N - x = 750 - 50 = 700 \, mm$$

$$Y = 11 \times 21.25 \times 700 / (11 \times 21.25 + 150) = 426 \, mm$$

Base Plate sizing
Taking moment about T,

$$C = \left[P.(d - N/2) + Mx \right] / (d - Y/3)$$

$$= \left[300000 \times (700 - 0.5 \times 750) + 150000000 \right] / (700 - 426/3)$$

$$= 443548 \, N$$

$$= 444 \, kN$$

Required width of base plate required = Br

$$C = 1/2.Y.fc.Br$$

$$Br = 2C/(Y.fc) = 2 \times 444000 / (426 \times 21.25) = 98 \, mm < B \text{ provided.}$$

Actual width of base plate shall be determined after selecting nos. and spacing of anchor bolt.
Anchor bolt

$$T = C - PT = 444 - 300 = 144 \, kN$$

$$\text{Bolt provided : 3 Nos 24 mm dia} \left(\phi_{dia} \right) \quad fy = 250 MPa$$

$$\text{Tensile capacity of one bolt} == 3.14 \times (0.87 \times 24)^2 / 4 \times 250 / 1000 = 86 kN$$

$$\text{Total tension capacity} = 3 \times 86 = 258 kN$$

$$\text{Factor of safety}, F = 258/144 = 1.79 < 1.67 ; \text{Safe.}$$

Minimum width of base plate to accommodate anchor bolts = 261 mm

$$3 \times 3 \times (24 + 5) = 261 \, \text{mm}$$

Width provided, B = 400 mm < 261 mm; Safe.

Base plate sizing
Length, N = 750 mm; Width, B = 400 mm
C = 444 kN Y = 426 mm

$$\text{Avg.base pressure,} p = 444000 / (400 \times 426 / 2) = 5.21 \text{MPa}$$

$$C = 1/2 \, Y.B.p \qquad p = C / (B.Y / 2)$$

$$\text{max.pressure at edge of base plate,} p = 5.21 \times 2 = \mathbf{10.42} \text{MPa}$$

Taking section at outside face of main gusset welded to column flange
Stiffener – b
Height = 250 mm
Thickness = 12 mm
Base plate
Width = 400 mm
Thickness = 12 mm

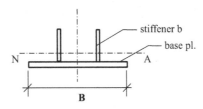

FIGURE 3.29   Sectional view.

NA dist. from bottom of base plate:

$$y = \left[2 \times 250 \times 12 \times (125 + 25) + 400 \times 25 \times 12.5 \right] / (2 \times 250 \times 12 + 400 \times 25) = 64 \, \text{mm}$$

$$I_{NA} = 2 \left[12 \times 250^3 / 12 + 250 \times 12 (125 + 25 - 64)^2 \right]$$

$$+ 400 \times 25^3 / 12 + 400 \times 25^3 / 12 + 400 \times 25 \times (64 - 0.5 \times 25)^2$$

$$= 102669333 \text{mm}^4$$

$$\text{Zx top} = 102669333 / (250 + 25 - 64) = 486585 \text{mm}^3$$

$$\text{Zx bot} = 102669333 / 64 = 1604208 \text{mm}^3$$

Resulting stress from base pressure, p

$$\text{Projection outside main gusset} = (750 - 450 - 2 \times 12) / 2 = 138 \text{mm}$$

$$M_1 = 10.42 \times 138^2 / 2 = 99219 \text{ Nmm} = \text{Upward moment from base pressure.}$$

$$\text{Comp stress at top} = 99219 / 486585 = 0.204 \text{MPa}$$

$$\text{Tensile stress at bot} = 99219 / 1604208 = 0.062 \text{MPa}$$

$$\text{Permissible stress} = 0.66 \times 165 \text{MPa}$$

Resulting stress from anchor bolt tension, T
Projection outside main gusset = 138 − 50 = 88 mm.

$$T = 144 \text{ kN}$$

$$M_2 = 144 \times 1000 \times 88 = 12672 \text{Nmm (tension at top)}$$

Tensile stress at top = 12672/486585 = 0.026 MPa
Comp stress at bot = 12672/1604208 = 0.008 MPa
Permissible stress = 0.6 × 250 = 150 MPa
Local stress from anchor bolt tension, T

$$\text{Force per bolt} = 144 / 3 = 48 \text{ kN}$$

Each bolt panel is having three end fixed and one end free, so the bolt tension will be resisted by three faces.

$$\text{Projection from stiffener b} = (133 - 12) / 2 = 60.5 \text{mm}$$

$$\text{M per face} = (48000 / 3) \times 60.5 = 968000 \text{N mm}$$

$$\text{Eff.width} = 60.5 \times 2 = 121 \text{mm}$$

$$Z = bt^2 / 6 = 121 \times 25^2 / 6 = 12604 \text{mm}^3$$

$$\text{Stress developed} = 968000 / 12604 = \textbf{77} \text{MPa}$$

$$\text{Permissible stress} = 0.66 \times 250 = \textbf{150} \text{MPa} > 77 \text{MPa; Safe.}$$

Stiffener – a

T = 144 kN Total bolt tension

Projection outside column face = (750 – 450)/2 = 150 mm

Avg. base pressure = 5.21 MPa B = 400 mm

$$C = \text{Total upward base pressure outside column flange}$$

$$= 400 \times 150 \times 5.21 / 1000 = 313 \text{kN}$$

Max. load transferred by main gusset Stiffener mkd – a = 313 kN

$$\text{Weld size}, \omega = 8\text{mm};$$

$$\text{fillet weld capacity} = 0.7 \times 8 \times 110 = 616 \text{N / mm}$$

No. of vertical weld run = 2

$$\text{Length of weld required / side} = 312600 / (2 \times 616) = 254 \text{mm}$$

Height of stiffener – a = 350 mm. Safe

Stiffener – b included in base plate design

Shear lug

Shear force, Fx = 30 kNfc' = 25 MPa

Lug: Embed. depth below grout, Le = 125 – 40 = 85 mm (Le = L – G)

Width = 150 mm

Following criteria are used to determine embedded length and shear capacity of lug:

(ACI 318-19 Table 22.8.3.2)

(i)   Bearing strength of concrete

(ii) Shear strength of concrete in front of lug

(iii) Strength design

Bearing capacity of concrete:

Bn = 0.85 fc'A1, where, A1 = bearing area below grout

$$= 0.85 \times 25 \times 85 \times 150 / 1000 = 271 \text{kN}$$

$$\text{Bu} = 271 / 1.67 = 162 \text{kN} > \text{Fx; Safe.}$$

Shear strength of concrete in front of lug, Fr:

$$\text{Projected area of the failure plane at the face of pedestal} = a \times b$$

$$\text{Vertical length} = 85 + 475 = 560 \text{mm}$$

$$\text{Horizontal width} = \text{Wp} = \text{pedestal width} = 600 \text{mm}$$

$$\text{Effective stress area} = 560 \times 600 - 85 \times 150 = 323250 \text{mm}^2$$

Uniform tensile stress on eff. area $= 4\phi \sqrt{fc} = 4 \times 0.75 \times \sqrt{25} = 15 MPa$

$$Fr = 323250 \times 15 / 1000 = 4849 kN$$

$$Fr / \Omega = 4849 / 1.67 = 2904 kN > Fx; Safe.$$

(Note: The concrete design shear strength for the lugs shall be determined based on a uniform tensile stress of $4\phi \sqrt{fc'}$ acting on an effective stress area defined by projecting a 45-degree plane from the bearing edge of the shear lug to the free surface. The bearing area of the lug is to be excluded from the projected area. Use $\phi = 0.75$. Refer ACI Steel design guide 1 for base plate design / App B of ACI 349-01.)

Strength design:

$$Mu = Fx(G + Le/2) kNm = 30 \times (40 + 85/2) / 1000 = 2.475 \, kNm$$

$$Mn = Fy Zp / \Omega = Fy(bt_1^2 / 4) / \Omega = Mu$$

$$t_1 = \sqrt{[4M\Omega / (Fy.b)]} = \sqrt{[4 \times 2475000 \times 1.67 / (250 \times 150)]} = 21mm$$

Thickness provided $= 25mm > t_1 = 21mm; Safe.$

(Alternately, shear key with two flats making a cross pattern may be provided to reduce plate thickness.)

3.6.2 Twin Legged Column – IS 800

a) Materials

Concrete grade $= M25$

Characteristic strength, $fck = 25 N / mm^2$

Bearing pressure on concrete, $\sigma_{cbc} = 8.5 N / mm^2$

Modulus of elasticity, $Ec = 28500 N / mm^2$

Structural steel and anchor bolt: IS 2062 Grade E 250 (Fe 410 W) A

Yield stress, $fy = 230 N / mm^2$

Modulus of elasticity, $Es = 200000 N / mm^2$

b) Design Load

TABLE 3.12
Column Base Loads

		Axial	Moment	Shear	
		P	Mx	Sy	
Load Case	Description	kN	kN	kN	Load Factor
1	Dl +LLR	3500	800	450	1
2	DL + LL+EQP	4900	1000	550	1
3	0.75(DL+WL)	1750	1500	625	1
4	0.75(DL+%LL+EQP+WL)	4300	1200	750	1
5	0.75(0.9 DL +WL)	1250	1600	600	1

Note: DL = Dead Load; LLR = Roof Live Load; LL = Live Load on Floors; EQP = Equipment Load; WL = Wind Load.

c) Dimensions

Column size: Twin legged column
   Column size, d = 1500 mm
   Depth of Col leg, D = 750 mm
   Width of Col leg, W = 400 mm
   Flange thickness, tf = 25 mm
   Web thickness, tw = 16 mm

Base Plate:
   Length, L = 2100 mm
   Width, B = 1050 mm
   Thickness, ts = 25 mm

Anchor Bolt:
Nos. per side, n = 4

$$\text{Nominal dia.} \phi_{dia} = 28 \text{mm}$$

$$\text{Net area} \left(0.7 \text{ gross area} \right), \phi_{area} = 431 \text{mm}^2$$

$$\left(0.7 \pi \phi_{dia}^2 / 4 \right)$$

$$\text{Bolt pretension stress}, \text{fpt} = 126 \text{N} / \text{mm}^2$$

$$\text{Edge distance of bolt holes}, \text{e} = 75 \text{mm}$$

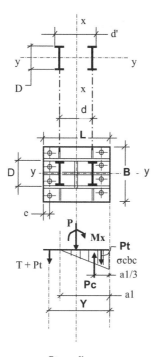

Stress diagram

FIGURE 3.30   Twin legged Column base.

Lever arm for moment capacity, Y

$$Y = L - e = 2025\,\text{mm}$$

Gusset plate: (IS800 :2007 – 7.7.2.3)
   Main gusset: h1 = 1500 mm; t1 = 20 mm
   Aux. gusset: h1 = 300 mm; t2 = 12 mm
   Pt = Pretension on bolt; T = Tension in bolt.

d) To Determine Bearing Pressure and Moment Capacity:

Pretension in anchor bolts, $Pt = n\,.fpt\,.\phi_{\text{area}} = 4 \times 126 \times 431 = 217\,\text{kN}$

   To find out NA depth, a1 with the conventional triangular distribution, we get

$$2.Pt + P = B.\sigma_{\text{cbc}}.a1\,/\,2, \text{where } a1 = 2Pt + P\,/\,(0.5B \times \sigma_{\text{cbc}})$$

Load Case 1

$$P = 3500\,\text{kN}; Mx = 800\,\text{kN}$$

$$B = 1050\,\text{mm}; L = 2100\,\text{mm}$$

$$\sigma_{cbc} = 8.5 M / mm^2; e = 75 mm$$

$$al = 2 \times 21700 + 3500000 / (0.5 \times 1050 \times 8.5) = 882 mm$$

$$Pc = 0.5.B.\sigma_{cbc}.al = 0.5 \times 1050 \times 8.5 \times 882 / 1000 = 3936 kN$$

$$Y = L - e = 2100 - 75 = 2025 mm$$

$$Lever\,arm, X = Y - al / 3 = 2025 - 882 / 3 = 1731 mm$$

$$Moment\,capacity = Pc.Lever\,arm\,(X) = 3936 \times 1731 / 1000$$
$$= 6813\,kNm > Mx; Safe.$$

$$max.base\,pressure, pmax = P / LB + Mx / (BL^2 / 6)$$

$$= 3500000 / (2100 \times 1050) + 800000000 / (1050 \times 2100^2 / 6)$$

$$= 1.5873 + 1.0366 = 2.624 MPa < 8.5 MPa; Safe.$$

Permissible bearing pressure on pedestal = 8.5 MPa

TABLE 3.13
Summary of Bearing Pressure all the Load Cases

Load Case	Axial P kN	Moment Mx kNm	Base Pressure p_{max} Mpa	Base Pressure p_{min} MPa	Triangular Stress Block a1 mm	Triangular Stress Block X mm	Triangular Stress Block Mcap kNm	Observation
1	3500	800	2.62	0.55	882	1731	6811	Safe
2	4900	1000	**3.52**	0.93	1195	1627	8677	Safe
3	1750	1500	2.74	−1.15	490	1862	4067	Safe
4	4300	1200	3.51	0.40	1061	1671	7913	Safe
5	1250	1600	2.64	−1.51	377	1899	3199	Safe

---

e) **Base Plate Thickness, ts**

$$ts = \sqrt{(2.5 wc^2 \gamma_{mo} / fy)} > tf\,(IS\,800 : 2007 - 7.4.1.1)$$

where,

Uniform base pr, $w = p_{max} / 2 = 1.76 Mpa$; $p_{max} = 3.52\,MPa$

Projection beyond main gusset, $c = 130\,mm$

$$c = \left(1050 - 750 - 2 \times 20\right) / 2 = 130mm$$

Partial safety factor for material, $\gamma mo = 1.10\left(IS\,800:2007 - Table\,5\right)$

$$fy = 230\,Mpa$$

$$ts = \sqrt{\left(2.5 \times 1.76 \times 130^2 \times 1.1 / 230\right)} = 20mm < ts\,Provided = 25mm; Safe.$$

Tension in a single bolt, Pt = 431 × 126/1000 = 54 kN
    Panel dimension:

$$a = 1050 / 4 = 263\,mm$$

$$b = \left(2100 - 1500 - 16\right) / 2 = 292mm$$

Bolt tension will have three edge supports.

$$T\,per / support\,side = 18\,kN$$

$$M = T \times a / 2 = 18000 \times 263 / 2 = 2367000Nmm$$

$$Eff.width\,at\,support = 2\left(a / 2\right) = 263\,mm$$

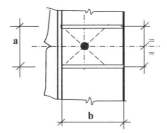

FIGURE 3.31   Part Plan

Thickness = 25 mm

$$Z = b\,t^2 / 6 = 263 \times 25^2 / 6 = 27396mm^3$$

$$Stress\,developed = M / Z = 2367000 / 27396 = 86MPa$$

Permissible bending stress $= 0.75\,fy = 0.75 \times 230$
$$= 172.5\,MPa; Safe.\left(IS:800:2007 - 11.3.3\right)$$

f) Gusset Plate

max .pressure at edge of base plate, $p_{max} = 3.52$ Mpa

Taking section through main gusset at outer edge of column flange

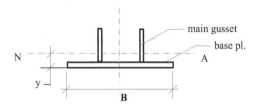

FIGURE 3.32   Sectional view.

Main Gusset
Height = 1500 mm; Thickness = 20 mm
[Aux gusset ignored]
Base Plate
Width = 1050 mm; Thickness = 25 mm
NA dist. from bottom of base plate: y = 543 mm

$$I_{NA} = 21868351250 \text{mm}^4$$

$$Zx \text{top} = 21868351250 / (1500 + 25 - 543) = 22269197 \text{mm}^3$$

$$Zx \text{bot} = 21868351250 / 543) = 40273207 \text{mm}^3$$

$$\text{Projection outside column flange} = (2100 - 1500 - 400) / 2 = 100 \text{mm}$$

$$M_1 = 3.52 \times 100^2 / 2 = 17600 \text{Nmm} = \text{Upward moment from base pressure}$$

$$\text{Comp stress at top} = 17600 / 22269197 = 0.001 \text{MPa}$$

$$\text{Tensile stress at bot} = 17600 / 40273207 = 0.0004 \text{MPa}$$

$$\text{Permissible stress} = 0.75 \times 230 = 172.5 \text{MPa} \left(\text{IS} : 800 : 2007 - 11.3.3 \right)$$

Safe.
    Welding:
    Horz: Base Plate to Gusset – 10 mm; Fillet weld all round
    Vert: Main gussets to column shaft – 8 mm; Fillet weld both side
    Aux. gussets to column/main gusset – 6 mm; Fillet weld both side
Shear Lug: Method of design same as done in 3.6.1.

REFERENCES

1. Steel Construction Manual AISC – Fourteenth edition.
2. IS 800: 2007 General Construction in Steel – Code of Practice.
3. AISC 360 – 10 Specification for Structural Steel Buildings.
4. Design of Welded Structure – Omer W. Blodgett.

4 Bolted Joints
Workout Examples

4.1 BOLT VALUE OR DESIGN STRENGTH OF BOLTS

Bolts are subject to shear and externally applied tension. The bolt value or design strength of a bolt is defined as the product of "the effective area of bolt" and "the permissible stress." The bolt value is used to determine the number of bolts in a joint (Number of bolts in a joint = Applied force/Bolt value). In this chapter, we will discuss how to calculate the permissible stress and bolt value according to AISC specification and Indian codes.

a) **Reference**
   1. AISC Steel construction manual – 14 Edition
   2. IS 800:2007/IS 816:1969/IS 1367:1967
b) **Material**

TABLE 4.1
Properties of Bolts and Connection Materials

Plates, Angles, and Connection Materials

Yield stress	fy	36	ksi	250	MPa
Tensile stress	Fu	65	ksi	410	MPa
Elasticity	E	29000	ksi	200000	MPa
Bolts					
HT Bolt 8.8 ($d \leq 16$)	Fyb	640	Fub	800	MPa
HT Bolt 8.8 ($d > 16$)	Fyb	660	Fub	830	MPa
ASTM A325					
(Diameter 24 mm and less)	Fyb	640	Fub	800	MPa
(24 mm > dia > 37 mm)	Fyb	660	Fub	830	MPa
ASTM A490 ($d <= 37$)		150	ksi	1034	MPa

 DOI: 10.1201/9781003539124-4

c) Sketch

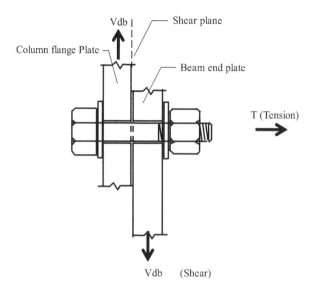

FIGURE 4.1   Forces in a bolt.

4.1.1   BEARING TYPE BOLT – AISC (ASD)

The bearing type bolts carry tension and shear force. The tension capacity of a bolt depends on the net area of the bolt shank at root excluding threads. The shear capacity of a bolt is the smaller of shear strength of the shank and bearing strength on its contact area over connecting plates. The following example will show how to calculate the strength of bolts.

Bolt: M24 (A325X)

[X = threads Excluded from shear plane; N = threads Not excluded from shear plane.]
d, bolt dia. = 24 mm; d_0 = dia. of hole = 27 mm (AISC – Table J3.3M; J3.2)

Nominal tensile strength, Fnt = 620 MPa
Nominal shear strength, Fnv = 469 MPa

a) Allowable Shear Strength, Pv = Rn/Ω

$$Rn = Fnv.Ab \left(AISC\ manual\ J3-1 \right)$$

$$Fnv = 469\ MPa; \Omega = 2$$

$$Ab = 453\ mm^2 \left(area\ of\ bolt\ in\ unthreaded\ body\ area \right)$$

$$Bolt\ value\ in\ shear, Pv = 469 \times 453 / 2 / 1000 = 106\ kN$$

b) Bearing Strength at Bolt Holes, Pb = Rn/Ω

$$Rn = 1.2 lctFu \leq 2.4 dtFu \left(\text{AISC manual J3} - 6a \right)$$

where,
   Fu = specified tensile strength = 410 MPa
   d = nominal bolt diameter = 24 mm
   d_0 = diameter of bolt hole = 27 mm

Refer Table 4.5 below.

   lc = clear distance between edge of hole and the edge of adjacent hole or
       edge of the material
   = 30.5 mm $(e_0 - 0.5\ d_0)$ (44 – 0.5 × 27) (Refer Table 4.5 for lc)
   Ω = 2
   t = thickness of connecting material = 12 mm

$$1.2 lctFu = 1.2 \times 30.5 \times 12 \times 410 / 1000 = 180\ kN$$

$$2.4 dtFu = 2.4 \times 24 \times 12 \times 410 / 1000 = 283\ kN; Rn = 180\ kN$$

Bolt value in bearing, Pb = Rn / Ω = 180 / 2 = 90 kN

*Effective strength of the bolt or Bolt value is the lesser of fastener shear
strength or the bearing strength at the bolt hole.*
So, the bolt value, P = 90 kN
(Note: Block shear checking to be done for specific connections.)
Bolt: A325X AISC(ASD) Ω = 2.
How to find out bolt value from the chart:

Case I – Single shear and bearing on 12 mm plate
   Single shear value of 20 dia. bolt = 74 kN; Bearing Strength = 68 kN.
   Hence Design Strength of bolt will be 68 kN

Case II – double shear and bearing on 16 mm plate
   Double shear value of 20 dia. bolt = 74 × 2 = 148 kN; Bearing Strength
   = 91 kN.
   Hence Design Strength of bolt will be 91 kN
c) Allowable Tensile Strength, Pt = Rn/Ω

$$Rn = Fnt.Ab; Fnt = 620\ MPa; Ω = 2$$

$$Ab = 353\ mm^2 \left(\text{area of bolt in threaded part of the body} \right)$$

Bolt value in tension, Pt = 620 × 353 / 2 / 1000 = 109 kN

TABLE 4.2
Bolt Value or Strength of Bolts in Shear and Bearing, P(= Rn/Ω)

Bolt Dia.	Gross Area	Net Area	Hole Dia.	Single Shear[a]	Bearing on Plates of Thickness (mm)[b]							
					6	8	10	12	16	20	25	28
mm	mm²	mm²	mm	kN	kN	kN	kN	kN	kN	kN	kN	kN
16	201	157	18	47	32	43	54	65	86	108	135	151
18	255	192	20	60	36	48	59	71	95	119	122	166
20	314	245	22	**74**	34	45	57	**68**	**91**	113	135	159
22	380	303	24	89	42	56	70	84	112	140	149	197
24	453	353	27	106	49	66	82	99	132	165	162	231
27	573	459	30	134	58	78	97	117	156	194	182	272
30	707	561	33	166	66	87	109	131	175	219	203	306
33	856	694	697	201	63	84	105	126	168	211	223	295
36	1018	817	820	239	66	87	109	131	175	219	243	306

Notes:
a Fnv = 469 MPa.
b Fu = 450MPa.

4.1.2 SLIP-CRITICAL BOLT – AISC (ASD)

Slip-critical connections shall be designed to prevent slip and for the limit states of bearing-type connections.

Bolt: M24 (A490M)

a) Allowable shear, $Psc = Rn/\Omega = \mu D_u h_f T_b n_s$

where,
$\Omega = 1.5$ (ASD)
$\mu = 0.3$ Class a (unpainted clean mill scale)
$D_u = 1.13$ a multiplier factor
$T_b = 257$ kN bolt pretension
$h_f = 1$ factor for filler (1 for no filler, 0.85 for two or more filler)
$n_s = 1$ number of slip planes
$Rn = 0.3 \times 1.13 \times 1 \times 257 \times 1 = 87.12$ kN
$Psc = Rn/\Omega = 87.12/1.5 = \mathbf{58}$ kN (Single shear)

(Design strength of shear in double shear plane, Psc = 116 kN, for n = 2)

b) Slip resistance for bearing, $Pb = Rn/\Omega$
   $d = 24$ mm; $t = \mathbf{12}$ mm; $\Omega = 2$; $lc = 30.5$ mm

$$1.2\, lc\, t\, Fu = 1.2 \times 30.5 \times 400 = 180 \text{ kN}$$

$$2.4\, d\, t\, Fu = 2.4 \times 24 \times 12 \times 410 / 1000 = 283 \text{ kN}; Rn = 180 \text{ kN}$$

$$Pb = Rn / \Omega = \mathbf{90} \text{ kN}$$

Design strength of shear in single shear plane = **58 kN**
[Bolt value, Rn will be Psc or Pb, whichever is less.]

4.1.3   BEARING TYPE BOLT – IS 800

a) **In Limit State Design** [IS 800:2007 – 10.3.2]
   Assumptions: Threads excluded from shear planes.
   Bolt: M24 – 8.8 Grade

$$Fy = 640 \text{ MPa} \left(d \leq 16 \right) \text{and } 660 \text{ MPa} (d > 16) \left[\text{IS } 800 : 2007 - \text{Table } 1 \right]$$

$$Fu = 800 \text{ MPa} \left(d \leq 16 \right) \text{and } 830 \text{ MPa} (d > 16)$$

To determine design strength of Bolt, Vdsb:
   Vsb = Factored shear force; Vdb = Design strength of bolt; Vsb ≤ Vdb
   Vdb will be smaller of the Shear capacity of bolt (Vdsb) and Bearing capacity of bolt (Vdpb).
   Shear Capacity, Vdsb:

$$Vdsb = Vnsb / \gamma mb$$

where,
   Vnsb = Nominal shear capacity of bolt.
   γ_{mb} = Partial safety factor of bearing type bolt = 1.25
   Vnsb = (Fub/ $\sqrt{3}$) × (n_n. Anb + n_s. Asb) (IS 800:2007 – 10.3.3)

where,
   Fub = Ultimate tensile stress of bolt.
   n_n = Numbers of shear planes with threads intercepting the shear plane.
   n_s = Numbers of shear planes without threads intercepting the shear plane.
   Asb = Nominal plain shank area of the bolt (Gross area).
   Anb = Net shear area of the bolt at thread (area corresponding to root dia at thread).

Shear capacity of a bolt:

$$\text{Bolt diameter}, d = 24\,\text{mm}$$

$$\gamma_{mb} = 1.25;\ \text{Fub} = 830\,\text{MPa};\ n_n = 0;\ n_s = 1$$

$$\text{Asb} = 453\,\text{mm}^2;\ \text{Anb} = 353\,\text{mm}^2;\ \left[\text{Refer Table 4.5}\right]$$

$$\text{Vnsb} = \left[(830/1.732)\times(1\times453)\right]/1000 = 217\,\text{kN}$$

$$\text{Vdsb} = 217/1.25 = 174\,\text{kN}$$

$$\text{Bearing Capacity}, \text{Vdpb}$$

$$\text{Vdpb} = \text{Vnpb}/\gamma mb$$

$$\text{Vnpb} = \text{nominal bearing strength of a bolt} = 2.5\,\text{kb.d.t.Fu}$$

kb is smaller of $e_1/3d_0$, $(p/3d_0) - 0.25$, Refer Table 4.4.

where,
   e_1 = end distance along bearing direction
   p = pitch distance along bearing direction
   d_0 = diameter of the hole
   Fub = ultimate tensile stress of the bolt
   Fu = ultimate tensile stress of plate
   d = nominal diameter of the bolt
   t = summation of thicknesses of the connected plate experiencing bear-
      ing stress

Bearing Capacity of a bolt:

Bolt dia., d = 24 mm	e_1 = 44 mm	(refer Table 4.5)
d_0 = 27 mm	p = 60 mm	(2.5d min)
Fub = 830MPa	Fu = 450 MPa	

t = 20 mm (smaller of connecting flange of plate thickness, as applicable)

$$\gamma_{mb} = 1.25;\ \text{kb} = 0.491$$

$$\text{Bearing area}, \text{Apb} = 540\,\text{mm}^2\,(d\times t)$$

$$\text{Vnpb} = 2.5\times0.491\times24\times20\times450/1000 = 265\,\text{kN}$$

$$\text{Vdpb} = 265 / 1.25 = \textbf{212 kN}$$

$$\text{Vdb} = \text{smaller of Vdsb and Vdpb};\ \textbf{Vdb} = \textbf{174 kN}$$

Tension Capacity, Tb:

$$\text{Tb} \leq \text{Tdb}$$

where,
Tb = factored tensile force [IS 800:2007 CL 10.3.5]
$\text{Td} = \text{Tnb} / \gamma_{mb}$
Tnb = 0.9. Fub. An < Fyb. Asb. $(\gamma_{mb}/\gamma_{m0})$ [Nominal Tension capacity]

Bolt dia, d = 24 mm, where Fub = 830 MPa; $\gamma_{mb} = 1.25$; $\gamma_{m0} = 1.1$

$$\text{Fyb} = 660\text{MPa}; \text{An} = 353\,\text{mm}^2\,(\text{net area}); \text{Asb} = 453\,\text{mm}^2\,(\text{gross area})$$

$$0.9\,\text{Fub An} = 0.9 \times 830 \times 353 / 1000 = 264\ \text{kN}$$

$$\text{Fyb.Asb.}\left(\gamma_{mb}\,/\,\gamma_{m0}\right) = 660 \times 453 \times \left(125\,/\,1.1\right) / 1000 = 340\ \text{kN}$$

$$\text{Tnb} = 264\ \text{kN}$$

b) **Working Stress Design** (IS 800:2007 – 11.6.1)
Shear Capacity:
Permissible stress in bolt in shear, fasb

$$= 0.6\ \text{Vnsb} / \text{Asb} = \left(0.6 \times 217 / 453\right) \times 1000 = 287\,\text{MPa}$$

$$\text{Bolt value, Vsb} = \text{Asb} \times \text{fasb} = 453 \times 287 = 130\ \text{kN}$$

Bearing Capacity:
Perm. bearing stress in bolt/plate, fapb = 0.6 Vnpb/Apb = (0.6 × 265/540) × 1000 = 294 MPa
Bolt value, Vsb = Apb.fapb = $540 \times 294 / 1000 = 159\ \text{kN}$
Bolt value or design strength in shear, **Vsb = 130 kN**
Tensile Capacity:
Permissible tensile stress of the bolt, fatb = 0.6 Tnb/An = (0.6 × 264/353) × 1000 = 449 MPa
Bolt value in tension = An.fatb = $353 \times 449 = 158\ \text{kN}$
Bolt value or Design tension capacity, **Td = 158 kN**

TABLE 4.3

Bolt Value or Strength of High Tensile 8.8 Grade Bolts in Shear and Bearing

Bolt Dia.	Gross Area	Net Area	Hole Dia.	Single Shear[a]	Bearing on Plates of Thickness (mm)[b]							
					6	8	10	12	16	20	25	28
mm	mm²	mm²	mm	kN	kN	kN	kN	kN	kN	kN	kN	kN
16	201	157	18	56	32	42	53	64	85	106	133	148
18	255	192	20	73	36	49	61	73	97	122	152	170
20	314	245	22	90	39	52	65	79	105	131	164	183
22	380	303	24	109	46	61	76	92	122	153	191	214
24	453	353	27	130	48	64	80	95	127	159	199	223
27	573	459	30	165	55	73	91	109	146	182	228	255
30	707	561	33	203	62	82	103	123	165	206	257	288
33	856	694	697	246	69	92	114	137	183	229	286	321
36	1018	817	820	293	75	100	125	150	199	249	312	349

Notes:

[a] $V_{sb} = 0.6(f_u/\sqrt{3}).A_{sb}$.

[b] $V_{sb} = 0.6(2.5 k_b d t f_u)$.

(IS 800:2007 Working stress design.)

From the above table, we can get the bolt values directly for design of joints.

For example, see the joint detail shown in Figure 4.1. The design load is 150 kN (V_{db}), and the thickness of connected plates are 12 mm each.

Let us choose M24 bolt (8.8 grade) for the connection. From the above bolt value table, the single shear capacity of a M24 bolt is 130 kN and bearing capacity on 12 mm plate is 95 kN. So, the bolt value of will be lower of the two, i.e., 95 kN.

So, let us provide two bolts: Design capacity = 2 × 95 kN = 190 kN > 150 kN design load;

Hence, Safe.

4.1.4   FRICTION GRIP BOLT (IS 800:2007 – 11.6.1)

Let us use M24 HT Bolt 8.8 grade – friction grip.

Bolt diameter, d = 24 mm

γ_{mb} = 1.25 Material safety factor.

F_{ub} = 830 MPa Ultimate tensile stress of bolt.

F_u = 410 MPa Ultimate tensile stress of plate.

n_0 = 0 No. of shear planes with threads intercepting the shear plane.

n_s = 1 No. of shear planes without threads intercepting the shear plane.

A_{sb} = 453 mm² Nominal plain shank area of the bolt.

Anb = 353 mm^2 Net shear area of the bolt at thread.

e_1 = 44 mm End distance along bearing direction.

d_0 = 27 mm Diameter of the hole.

p = 72 mm Pitch distance along bearing direction (2.5d min).

kb = 0.543 [kb = smaller of e/3d0, (p/3d0 – 0.25), fub/fu, 1]; Refer Table 4.4.

i) **By working stress method: (IS 4000 1992)**

Slip-critical connections shall be designed to prevent slip and for the limit states of bearing-type connections.

Bolt: M24 HT Bolt 8.8 grade – friction grip

The slip resistance for the limit state of slip = Rn; Allowable Shear, Psc = Rn/f.o.s

$$Rn = \mu\, D_u h_f Tb\, n_s$$

where,

f.o.s = 1.4 Factor of Safety

μ = 0.35 Slip factor – Class a (unpainted clean mill scale)

Du = 1 A multiplier factor

Tb = 212 kN Bolt pretension

h_f = 1 Factor for filler (1 for no filler, 0.85 for two or more filler)

n_s = **1** Number of slip planes

$$Rn = 0.35 \times 1 \times 1 \times 212 \times 1 = 74.2\ kN$$

$$Psc = Rn\,/\,fos = 74.2\,/\,1.4 = 53\ kN\ single\ shear$$

Slip resistance for bearing, Vsb = 0.6 Vnpb

d = 24 mm; t = 20.8 mm (assumed); d_0 = 27 mm; p = 72 mm

e_0 = 44 mm; kb = 0.543

$$Vpnb = 2.5kbdtFu = 2.4 \times 0.543 \times 24 \times 20.8 \times 410\,/\,1000 = 267\ kN$$

$$\mathbf{Vsb} = 0.6 \times 267 = \mathbf{160}\ kN$$

Bolt value will be Psc or Vsb, whichever is smaller.

ii) **In Limit State: (IS 800:2007 10.4)**

Bolt: M24 HT Bolt 8.8 grade – friction grip

Slip resistance, Vsf = Vnsf = γ_{mf} where, γ_{mf} = 1.10 at service load and 1.25 at ultimate load

$$\mathbf{Vnsf = mf.n_e.Kh.F_0}$$

Where,

μf = 0.33 (coeff. of friction) (IS 800:2007 Table 20)

n_e = 1 (no. of slip surface); Kh = 1 (for fastener in clearance hole)

F_0 = 205 kN pretension in bolt

$\left(F0 = Anb.0.7fub = 353 \times 0.7 \times 830\,/\,1000 = 205\ kN\right)$

$$\text{Anb} = 353 \, \text{mm}^2$$
$$\text{Vdsf} = \left(0.33 \times 1 \times 1 \times 205\right)/1/1 = 62 \, \text{kN}$$

Bolt value will be Psc or Vsb, whichever is less.

iii) **Reference Tables**

TABLE 4.4
Value of Factor kb

kb (IS 800:2007 – 10.3.4)

	d	d_0	e_1	P	fub	fu	
Bolt size	mm	mm	mm	Mm	mm	mm	kb
M16	16	18	29	40	800	450	0.491
M18	18	20	32	45	830	450	0.5
M20	20	22	32	50	830	450	0.485
M22	22	24	38	55	830	450	0.514
M24	24	27	44	60	830	450	0.491
M27	27	30	51	67.5	830	450	0.5
M30	30	33	57	75	830	450	0.508
M33	33	36	57	82.5	830	450	0.514
M36	36	39	60	90	830	450	0.513

Note: kb = smaller of e/3d0, (p/3d0 – 0.25), fub/fu, 1.

TABLE 4.5
Properties of Bolts

	Bolt Dia.	Hole Dia.	Gross Area	Net Area	Edge Distance	
	d	d_0	Ag	An	e_1	lc
Bolt Size	mm	mm	mm²	mm²	mm	mm
M14	14	18	154	115	25	16
M16	16	18	201	157	29	20
M18	18	20	255	192	32	22
M20	20	22	314	245	32	21
M22	22	24	380	303	38	26
M24	24	27	453	353	44	30.5
M27	27	30	573	459	51	36
M30	30	33	707	561	57	40.5
M33	33	36	856	694	57	39
M36	36	39	1018	817	60	40.5

Note: Net area = area at threaded part.

4.2   BEAM TO BEAM – SHEAR CONNECTION

In this section, we will provide workout examples for connections of beams with web cleats and face plates.

4.2.1   WEB FRAMING CLEAT ANGLE – IS 800 (WORKING STRESS)

a) Sketch

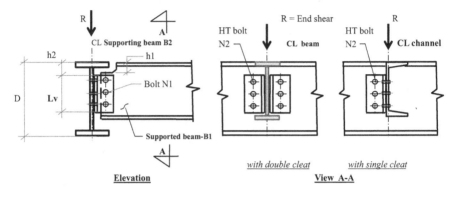

FIGURE 4.2   Shear connection with web cleats.

b) Materials

TABLE 4.6
Properties of Bolt & Structure – IS Standard

IS: 800 – 2007

Grade			E 250 (Fe 410 W) A			
Yield stress	fy	250	240	230		MPa
d or t		<20	20-40	>40		
Tensile Stress	fu	410	410	410		MPa
Elasticity	E	200000				MPa
Material Safety	γ_{m0}	1.10	γ_{m1}	1.25		for plate
Factor		Bolt friction type		Bolt friction type		
	γ_{mf}	1.25	γ_{mb}	1.25		
HT Bolt IS:1367						
8.8 (d ≤ 16)	Fyb	640	Fub	800		MPa
8.8 (d > 16)	Fyb	660	Fub	830		MPa

---

c) Member Description

Supported member: B1	ISMB 250	D = 250 mm
Supporting member: B2	ISMB 250	D = 250 mm

d) Design Load, R

End Shear 70% of web shear capacity. (Assumed for Joint design)
Web shear:

Effective depth of web, $D - h1 = 222$ mm; $h1 = 28$ mm
Web thickness, $t = 6.9$ mm; $h2 = h1 + 15 = 43$ mm
Shear stress, $\tau sh = 0.4\,fy = 0.4 \times 250 = 100$ MPa (IS 800 : 2007 – 11.4.2; 8.4.1)

$$R = \text{End shear} = 0.70 \times (D - h1) \times t \times 0.4\,fy\,/1000$$

$$= 0.7 \times 222 \times 6.9 \times 0.4 \times 250\,/1000 = \mathbf{107}\,kN$$

e) Cleat sizing: Provide web cleat angles: **2 ISA 75 75 10**

Length of cleat, $Lv = 175$ mm
Leg thickness, $Lt = 10$ mm
Leg size, $Lh = 75$ mm

Shear capacity of cleats = $2 \times 175 \times 10 \times 100/1000 = 350kN > R$; Okay.

f) Design of connection bolts – Bearing Type

Let us use **M20 HT Bolt 8.8 grade – Bearing type bolt**

Bolt diameter, $d = 20$ mm
$\gamma_{mb} = 1.25$ Material Safety factor.
Fub = 830 MPa Ultimate tensile stress of bolt.
Fu = 410 MPa Ultimate tensile stress of plate.
$n_n = 0$ No. of shear planes with threads intercepting the shear plane.
$n_s = 1$ No. of shear planes without threads intercepting the shear plane.
Asb = 314 mm² Nominal plain shank area of the bolt.
Anb = 245 mm² Net shear area of the bolt at thread.
$e_1 = 32$ mm End distance along bearing direction.
$d_0 = 22$ mm Diameter of the hole.
$p = 50$ mm Pitch distance along bearing direction (2.5d min).
kb = 0.485 [kb = smaller of e/3d0, (p/3d0 – 0.25), fub/fu, 1]; Refer Table 4.4.

Shear capacity:

$$Vnsb = (Fub\,/\sqrt{3}) \times (n_n.Anb + n_sAsb\,(IS\,800:2007 - 10.3.3)$$

$$Vnsb = \left[\left(830/1.732 \right) \times \left(1 \times 314 \right) \right] / 1000 = 150 \, kN$$

Permissible stress in bolt in shear, fasb = 0.6 Vnsb / Asb

$$= \left(0.6 \times 150 / 314 \right) \times 1000$$
$$= 287 \, MPa. \left(IS\, 800 : 2007 - 11.6.2.1 \right)$$

Bolt value, Vsb = Asb × fasb = 314 × 287 = 90 kN (single shear)

Bearing capacity:
Bearing Plate thickness = 6.9 mm [smaller of beam web and cleat thickness].

Apb = 138 mm² bearing area on plate (d × t)

Vnpb = nominal bearing strength of *a* bolt = 2.5 kb. *d. t*. Fu (IS 800 – 2007 – 10.3.4)

$$Vpnb = 2.5 \times 0.485 \times 20 \times 6.9 \times 410 / 1000 = 69 \, kN$$

Perm. Bearing stress in bolt / plate, fapb = 0.6 Vnpb / Apb

=(0.6 × 69/138) × 1000 = 300 MPa (IS: 800 – 2007 – 11.6.2.2)
Bolt value, Vsb = Apb. fapb = 138 × 300/1000 = **41** kN
Bolt N1 End shear = **107** kN
Number of shear plane, $n = 2$
Shear Capacity = 2 × 90 = 180 kN; Bearing Capacity = 41 kN
Bolt value or design strength in shear, Vsb = 41 kN per bolt.
Number of bolt per side (N1) = 3

Total capacity = 3 × 41 = **123** kN > R = 107 kN; Safe

Bolt N2 End shear = **107** kN
Number of shear plane, n = 1
Shear Capacity = 1 × 90 = 90 kN; Bearing Capacity = 41 kN
Bolt value or design strength in shear, Vsb = 41 kN per bolt.
Number of bolt per side (N2) = 6

Total capacity = 6 × 41 = **246** kN > R = 107 kN; Safe.

g) **Block Shear Capacity, Tb; R = 107 kN**
   (V = Block shear strength; V1 = shear; T1 = tension)
   Web Cleat: 2 ISA 75 75 10
   t = 10 mm; b_L = bs = 75 mm; e_1 = 32 mm

Length of cleat, Lv = 175 mm

L = 175 – 32 = 143 mm; $b1$ = 32 mm

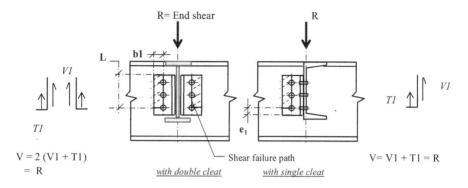

FIGURE 4.3   Block shear rupture.

Connecting Bolts: M20 8.8 grade
$d_0 = 22$mm; nos $= 3$
Tdb = Smaller of Tdb1 or Tdb2 (IS 800:2007 – 6.4.1; 6.4.2)

$$\textbf{\textit{Tdb1}} = \left[A_{vg}.fy / \left(\sqrt{3}\, \gamma_{mo} \right) + 0.9\, A_{tn}.fu \,/\, \gamma_{m1} \right]$$

Or

$$\textbf{\textit{Tdb2}} = \left(0.9 A_{vn}.fu \,/ \left(\sqrt{3}\gamma_{m1} \right) + A_{tg}\ fy \,/\, \gamma_{mo} \right)$$

Avg = gross area in shear along bolt line = $2 \times 143 \times 10 = 2860\,\text{mm}^2$

Avn = net area in shear along bolt line = $2860 - 2 \times 3 \times 22 \times 10 = 1540\,\text{mm}^2$

Atg = minimum gross area in tension perpendicular to the line of force = $2 \times 32 \times 10 = 640\ \text{mm}^2$

Atn = minimum net area in tension perpendicular to the line of force = $530\,\text{mm}^2$

$$\text{Tdb1} = \left[2860 \times 250 / \left(1.732 \times 1.1 \right) + 0.9 \times 530 \times 410 / 1.25 \right] / 1000\ \text{kN} = 532\ \text{kN}$$

$$\text{Tdb2} = \left[0.9 \times 1540 \times 410 / \left(1.732 \times 1.25 \right) + 640 \times 250 / 1.1 \right] / 1000\ \text{kN} = 408\ \text{kN}$$

$$\text{Tdb} = 408\ \text{kN}; \text{Tb} = 0.69\ \text{Tdb} = 281\ \text{kN}\left(\text{IS}\,800 : 2007 - 11.2.1\text{b} \right)$$

Block shear capacity, **Tb = 281 kN > R = 107 kN; Safe.**

4.2.2 WEB FRAMING CLEAT ANGLE – AISC (ASD)

a) Sketch
Refer Figure 4.2 Shear connection with angle cleats.
b) Material

TABLE 4.7
Properties of Bolt & Structure – ASTM

ASTM A 36		Table 2.4 AISC Manual			
Yield Stress	F_{nBM}	36	ksi	250	MPa
Tensile Stress	F_{uBM}	65	ksi	450	MPa
Elasticity	Es			200000	MPa
Electrode FE70		70	ksi		
Nominal Stress	F_{nw}	42	ksi	292	MPa
ASTM A325					
(dia. 24 mm and less)	Fu	120	ksi	827	MPa
(24 mm > dia. > 37 mm)		105	ksi	724	MPa
ASTM A490 ($d \le 37$)		150	ksi	1034	MPa

---

c) Member Descriptions

Supported member:	$W10 \times 17$	$D = 10.11$ inch
Supporting member:	$W10 \times 17$	$D = 10.11$ inch

d) Design Load, R
End Shear: 70% of web shear capacity. (Assumed for joint design)

$$\text{Effective depth of web}, D - h1 = 9.24 \text{ inch}; h1 = k = 0.875 \text{ inch}$$

$$\text{Shear stress}, \tau sh = 0.6 \text{ fy} = 0.6 \times 36 = 21.6 \text{ ksi}$$

$$\text{Web thickness}, tw = 0.24 \text{ inch}; h2 = h1 + 0.75 = 1.625 \text{ inch}$$

$$\text{Web shear capacity} = Rn = 0.6 \text{ fy Agv} = 0.6 \text{ fy} \left(D - h1\right) tw \left(\text{AISC} - \text{J4} - 3\right)$$

$$R = \text{Design end shear} = 0.70 \times \left(D - h1\right) \times tw \times 0.6 \text{fy} / \Omega$$

$$= 0.7 \times 9.24 \times 0.24 \times 0.6 \times 36 / 1.67 = \textbf{20 kips} \qquad \left(91 \text{ kN}\right)$$

e) **Cleat Sizing**: Provide web cleat angles: **2 L 2-1/2 × 2-1/2 × 5/16**

Length of cleat, Lv = 8 inch
Leg thickness, Lt = 5/16 inch (0.3125 inch)
Leg size, Lh = 2.5 inch
Shear capacity of cleats = 2 × 8 × 0.3125 × 21.6/1.67 = **65** kips > R;
  Okay.
Ω = 1.67 (AISC – G 2.1)

f) **Bolt Sizing**
  Bolt: A325X M16 Bearing Type

Bolt dia., d = 16 mm
dh = dia. of hole = 18 mm (AISC – Table J3.3M)
Nominal tensile strength, Fnt = 620 MPa (AISC – Table J3.2)
Nominal shear strength, Fnv = 469 MPa
[X = threads Excluded from shear plane; N = threads Not excluded from
  shear plane.]

Allowable Shear Strengh, Pv = Rn/Ω
Rn = Fnv. Ab (AISC manual J3-1)

$$Fnv = 469 \, MPa$$

Ab = 201 mm² (area of bolt in unthreaded body area)

$$\Omega = 2$$

Bolt value in shear, Pv = 469 × 201/2/1000 = **47** kN
Bearing Strength at Bolt Holes, Pb = Rn/Ω

$$Rn = 1.2 \, lc \, t \, Fu \leq 2.4 \, d \, t \, Fu \left(AISC \ manual \ J3-6a\right)$$

where,
  Fu = specified tensile strength = 827 MPa
  d = nominal bolt diameter = 16 mm
  lc = clear distance between edge of hole and the edge of adjacent hole
    or edge of the material = 20 mm; (edge dist. – 0.5dh) Refer Table 4.5.
  Ω = 2
  t = thickness of connecting material = 6.10 mm (minimum of web and
    cleat) (=0.24 inch)

$$1.2 \, lc \, t \, Fu = 1.2 \times 20 \times 6.1 \times 827 / 1000 = 121 \, kN$$

$$2.4dtFu = 2.4 \times 16 \times 6.1 \times 827 / 1000 = 194 \, kN; Rn = 121 \, kN$$

Bolt value in bearing, $Pb = Rn/\Omega = 121/2 = \mathbf{61}$ kN
Effective strength of the bolt or Bolt value is the lesser of fastener shear
strength or the bearing strength at the bolt hole.

Bolt N1: End Shear, $R = 20$ kips

Number of shear plane, n = 2
Shear Capacity = $2 \times 47 = 94$ kN; Bearing Capacity = 61 kN
Bolt value or Design strength in shear, Vsb = **61** kN per bolt.
Number of bolts per side (N1) = 3
Total Capacity = $3 \times 61 = 183$ kN = 40 kips > R; Safe.

Bolt N2: End Shear, $R = 20$ kips

Number of shear plane, n = 1
Shear Capacity = $1 \times 47 = 47$ kN; Bearing Capacity = 61 kN
Bolt value or Design strength in shear, Vsb = **47** kN per bolt.
Number of bolts per side (N2) = 6
Total Capacity = $6 \times 47 = 282$ kN = 62 kips > R; Safe.

g) **Block shear resistance, V = Rn/Ω**
   $R = 20$ kips; Refer Figure 4.3 above for block shear rapture diagram.
   V = Block shear strength; V1 = shear; T1 = tension
   Rn = 0.6 Fu Anv + Ubs Fu Ant < = 0.6 Fy Agv + Ubs Fu Ant (AISC J4-5)

$$\Omega = 2 \,(\text{ASD})$$

Web cleat: 2 L 2-1/2 × 2-1/2 × 5/16
Dimension of a single cleat
$e = 1.14$ inch; Lv = 8 inch; Lt = 5/16 inch; Lh = 2.5 inch; Cyy = 0.735
Bolt M16: nos/side = 3; hole dia. = 0.71 inch

$$L = 2 \times (8 - 1.14) = 13.72 \text{ inch}$$

$$b = 2 \times (2.5 - 0.735) = 3.53 \,\text{inch}; t = Lt = 5 / 16 \,\text{inch} \left(= 0.3125 \,\text{in}\right)$$

where,

$$Agv = \text{gross area subject to shear yielding} \left(L \times t\right)$$
$$= \left(13.72 \times 0.3125\right) = 4.29 \,\text{inch}^2$$

$$Anv = \text{net area subject to shear rupture} \left(Agv - \Sigma \text{hole area}\right)$$

$$= 4.29 - 2 \times 3 \times 0.71 \times 0.3125 = 2.96 \,\text{inch}^2$$

Ant = net area subject to tension $(b \times t) = 3.53 \times 0.3125 = 1.10 \, \text{inch}^2$

Ubs = 1 where tension is uniform; 0.5 where tension is not uniform.

Now,

$0.6 \, \text{Fu Anv} + \text{Ubs Fu Ant} = 0.6 \times 65 \times 2.96 + 1 \times 65 \times 1.10 = 187 \, \text{kips}$

$0.6 \, \text{Fy Agv} + \text{Ubs Fu Ant} = 0.6 \times 36 \times 4.29 + 1 \times 65 \times 1.10 = 164 \, \text{kips}$

$\text{Rn} = 164 \, \text{kips}; \text{V} = \text{Rn} / \Omega = 164 / 2 = 82 \, \text{kips}$

Block shear strength, V = 82 kips > R = 20 kips; Safe.

4.2.3   END CONNECTING FACE PLATE – IS 800 (WORKING STRESS)

a) Sketch

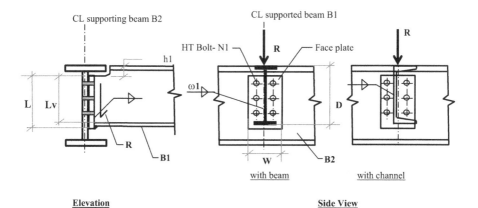

FIGURE 4.4   Shear connection with face plate – bolted.

b) Materials
Refer Table 4.6 Properties of Bolt & Structure – IS standard
c) Member Descriptions

Supported member: B1	ISMB 300	D = 300 mm
Supporting member: B2	ISMB 300	D = 300 mm

d) Design Load, R
End Shear 70% of web shear capacity. (Assumed for Joint design)
Effective depth of web, D − h1 = 271 mm; h1 = 29 mm
Web thickness, tw = 7.5 mm; Flange thickness, tf = 12.4 mm
Shear stress, τsh = 0.4 fy = 0.4 × 250 = 100 MPa (IS 800:2007 – 11.4.2;
8.4.1)

$$R = \text{End shear} = 0.70 \times (D - h1) \times tw \times 0.4\, fy / 1000$$
$$= 0.7 \times 271 \times 7.5 \times 0.4 \times 250 / 1000 = \mathbf{142} \text{ kN}$$

e) Face Plate Sizing

$$L \text{ provided} = D - h1 + 15\,mm = 285\,mm$$

Lv = Effective height of weld connecting web to face plate = 271 – 12.4 = 258.6 mm

W = Width of the plate = width of the flange of the supported beam, W = 140mm

Thickness of plate (~Thickness of web of supported beam, tp), t = 8 mm

f) Fillet Weld Sizing

Electrode: EX40XX; fy = 330 MPa

Permissible Stress in fillet weld, fw = 132 MPa (shop weld – 0.4 fy; IS 800 – 11.6.3)

ω_1 = fillet weld size = 6 mm (~thickness of web)

Strength of fillet welding connecting supported beam web to face plate,

$$F1 = 2Lv \times 0.7\omega_1 \times fw = 2 \times 258.6 \times 0.7 \times 6 \times 132 / 1000 = 287\,kN > R; \mathbf{Safe}$$

g) Design of Connection Bolts – Bearing Type

Let us use, **M16 HT Bolt 8.8 grade – Bearing type bolt**

Bolt diameter, d = 16 mm

γ_{mb} = 1.25 Material Safety factor

Fub = 800 MPa Ultimate tensile stress of bolt.

Fu = 410 MPa Ultimate tensile stress of plate.

n_0 = 0 Numbers of shear planes with threads intercepting the shear plane.

n_s = 1 Numbers of shear planes without threads intercepting the shear plane.

Asb = 201 mm^2 Nominal plain shank area of the bolt.

Anb = 157 mm^2 Net shear area of the bolt at thread.

e_1 = 29 mm End distance along bearing direction.

d_0 = 18 mm Diameter of the hole.

p = 40 mm Pitch distance along bearing direction (2.5d min).

kb = 0.491 [kb = smaller of e/3d0, (p/3d0 – 0.25), fub/fu, 1]; Refer Table 4.4.

Shear capacity

$$Vnsb = (Fub / \sqrt{3}) \times (n_n.Anb + n_s.Asb)(IS\,800 : 2007 - 10.3.3)$$

$$Vnsb = \left[(800 / 1.732) \times (1 \times 201) \right] / 1000 = 93\,kN$$

Permissible stress in bolt in shear,

$$\text{fasb} = \frac{0.6\,\text{Vnsb}}{\text{Asb}} = \left(0.6 \times \frac{93}{201}\right) \times 1000 = 278\,\text{MPa.}\left(\text{IS}\,800:2007-11.6.2.1\right)$$

$$\text{Bolt value, Vsb} = \text{Asb} \times \text{fasb} = 201 \times 278 / 1000 = 56\,\text{kN}\left(\text{Single shear}\right)$$

Bearing capacity
Bearing Plate thickness = **7.5** mm [smaller of beam web and face plate thickness].
  Apb = 120 mm² bearing area on plate (16 × 7.5)

$$\text{Vnpb} = \text{nominal bearing strength of a bolt}$$
$$= 2.5\,\text{kb.d.t.Fu}\left(\text{IS}\,800:2007-10.3.4\right)$$

$$\text{Vpnb} = 2.5 \times 0.491 \times 16 \times 7.5 \times 410 / 1000 = 60\,\text{kN}$$

Perm. Bearing stress in bolt/plate, fapb = 0.6 Vnpb/Apb = (0.6 × 60/120) × 1000 = 300 MPa (IS 800:2007 – 11.6.2.2)
  Bolt value, Vsb = Apb. fapb = 120 × 300/1000 = **36** kN
  Bolt N1 End shear = **142** kN
  Number of shear plane, n = **1**
  Shear capacity = 1 × 56= 56 kN; Bearing capacity = 36 kN
  Bolt value or Design strength in shear, Vsb = 36 kN per bolt.
  Number of bolt per side (N1) = 6

$$\text{Total capacity} = 6 \times 36 = 216\,\text{kN} > R = 142\,\text{kN; \textbf{Safe.}}$$

h) **Block shear capacity, Tb: R = 142 kN**

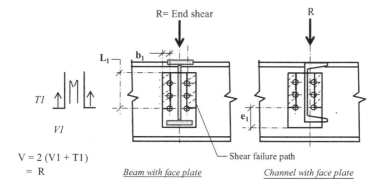

FIGURE 4.5 Block shear rupture.

Member section
ISMB 300; Flange thickness, tf = 12.4 mm
Face plate
Length, L = 285 mm; Width, W = 140 mm; Thick, Th = 8 mm
Connection bolts; **M16; 8.8 grade**
Bolt dia., d = 16 mm; Hole dia., d_0 = 18 mm; Nos = 2 × 3 = 6

$$e_0 = 29 \, \text{mm}$$

Now,

$$e_1 = e_0 + tf + 15 = 29 + 12.4 + 15 = 56.4 \, \text{mm}$$

$$L_1 = L - e_1 = 285 - 56.4 = 229 \, \text{mm}; b1 = 29 \, \text{mm}$$

Tdb = Smaller of Tdb1 or Tdb2 (IS 800:2007 – 6.4.1; 6.4.2)

$$\textbf{\textit{Tdb1}} = \left[A_{vg}.fy \, / \left(\sqrt{3} \, \gamma_{mo} \right) + 0.9 \, A_{tn}.fu \, / \, \gamma_{m1} \right]$$

Or

$$Tdb2 = \left[0.9 A_{vn}.fu \, / \left(\sqrt{3} \gamma_{m1} \right) + A_{tg} \, fy \, / \, \gamma_{mo} \right]$$

Avg = gross area in shear along bolt line = $2 \times 229 \times 8 = 3664 \, \text{mm}^2$

Avn = net area in shear along bolt line = $3664 - 2 \times 3 \times 18 \times 8 = 2800 \, \text{mm}^2$

Atg = minimum gross area in tension perpendicular to the line of force = 2 × 29 × 8 = 464 mm²

Atn = minimum net area in tension perpendicular to the line of force = $392 \, \text{mm}^2$

$$Tdb1 = \left[3664 \times 250 \, / \left(1.732 \times 1.1 \right) + 0.9 \times 392 \times 410 \, / \, 1.25 \right] / 1000 \, \text{kN} = 597 \, \text{kN}$$

$$Tdb2 = \left[0.9 \times 2800 \times 410 \, / \left(1.732 \times 1.25 \right) + 464 \times 250 \, / \, 1.1 \right] / 1000 \, \text{kN} = 583 \, \text{kN}$$

Tdb = **583** kN; **Tb = 0.69 Tdb = 402 kN** (IS 800:2007 – 7.5.1.2)
Block shear capacity, Tb = 402 kN > R = 142 kN; Safe.

4.2.4 END CONNECTING FACE PLATE – AISC (ASD)

a) **Sketch**
    Refer Figure 4.4
b) **Material**
    Refer Table 4.7

c) Member Descriptions

Supported member:	W10 × 17	D = 10.11 inch
Supporting member:	W10 × 17	D = 10.11 inch

d) Design Load, R
End Shear 70% of web shear capacity. (Assumed for Joint design)
Effective depth of web, D − h1 = 9.24 inch; h1 = k = 0.875 inch
Flange thickness, tf = 0.33 inch; Flange width, bf = 4.01 inch
Web thickness, tw = 0.24 inch

$$\text{Shear stress}, \tau sh = 0.6\,\text{fy} = 0.6 \times 36 = 21.6\,\text{ksi}$$

$$\text{Web shear capacity} = Rn = 0.6\text{fy Agv} = 0.6\,\text{fy}\,(D-h1)\,\text{tw}\,(AISC-J4-3)$$

$$R = \text{Design End shear} = 0.70 \times (D-h1) \times \text{tw} \times 0.6\,\text{fy}\,/\,\Omega;\ \Omega = 1.67$$

$$= 0.7 \times 9.24 \times 0.24 \times 0.6 \times 36 / 1.67 = \mathbf{20\,kips}\,(= 91\,kN)$$

e) Face Plate Sizing

$$\text{Lv required} = R\,/\,(\text{tw}.0.6\,\text{fy}) = 20\,/\,(0.24 \times 0.6 \times 36) = 4\,\text{inch}$$

L provided = D − h1 + 0.5 inch
L = 10.11 − 0.88 + 0.5 = 10 inch > Lv required: Okay.
W = Width of the plate = width of the flange of the supported beam
W = 4.01 inch; Provide 5-inch-wide face plate.
Thickness of plate ≥ Thickness of web of supported beam
tp = tw = 0.24 inch; Thickness provided = 5/16 inch (=0.313 inch)

f) Fillet Weld Sizing
Electrode Fe70XXX; fy = 70 ksi; Ω = 2
Permissible stress in fillet weld, fw = 21 ksi (shop weld;fw = 0.6fy/Ω)

$$\omega_1 = \text{Size of fillet weld provided} = 1\,/\,4\,\text{inch}$$

Strength of fillet welding connecting supported beam web to face plate,

$$F1 = 2.(D-\ h1-\ tf).0.707.\omega_1.fw + (2.bf-\ tw).0.707.\omega_1.fw$$

$$= 2 \times (10.11 - 0.88 - 0.33) \times 0.707 \times 0.25 \times 21$$
$$+ (2 \times 4.01 - 0.24) \times\ 0.707 \times 0.25 \times 21$$

$$= 65 + 29 = 94\,\text{kips} > R; \text{Safe}\,(AISC-J2-10a)$$

g) Bolt Sizing

Bolt: A325X M16 Bearing Type

Bolt dia., d = 16 mm

dh = dia. of hole = 18 mm (AISC – Table J3.3M)

Nominal tensile strength, Fnt = 620 MPa (AISC – Table J3.2)

Nominal shear strength, Fnv = 469 MPa

[X = threads Excluded from shear plane; N = threads Not excluded from shear plane.]

Allowable Shear Strengh, Pv = Rn/Ω

Rn = Fnv. Ab (AISC manual J3-1)

$$Fnv = 469 \, MPa$$

Ab = 201 mm^2 (area of bolt in unthreaded body area)

$$\Omega = 2$$

Bolt value in shear, Pv = 469 × 201/2/1000 = **47** kN **(10 kips)**

Bearing strength at Bolt Holes, Pb = Rn / Ω

$$Rn = 1.2 \, lc \, t \, Fu \leq 2.4 \, d \, t \, Fu \, (AISC \, manual \, J3 - 6a)$$

where,

Fu = specified tensile strength = 827 MPa

d = nominal bolt diameter = 16 mm

lc = clear distance between edge of hole and the edge of adjacent hole or edge of the material = 20 mm (edge dist. – 0.5dh) Refer Table 4.5

$\Omega = 2$

t = thickness of connecting material = 6 mm (minimum of web and cleat) (=0.24 inch)

$$1.2 \, lc \, t \, Fu = 1.2 \times 20 \times 6 \times 827 / 1000 = 119 \, kN$$

$$2.4 \, dt \, Fu = 2.4 \times 16 \times 6 \times 827 / 1000 = 191 \, kN; Rn = 119 \, kN$$

Bolt value in bearing, Pb = Rn / Ω = 119 / 2 = 59.5 kN $(13 \, kips)$

Effective strength of the bolt or Bolt value is the lesser of fastener shear strength or the bearing strength at the bolt hole.

Bolt N1 End Shear, R = 20 kips

Number of shear plane, n = **1**

Shear Capacity = 1× 47 = **47** kN; Bearing Capacity = **59.5** kN

Bolt value or Design strength in shear, Vsb = **47** kN per bolt.

Number of bolts per side (N1) = 6

Total Capacity = 6 × 47 = 282 kN = **62 kips > R = 20 kips; Safe.**

h) Block Shear Resistance, $V = Rn/\Omega$

R = 20 kips; Refer Figure 4.5 above for block shear rapture diagram.

V = Block shear strength; V1 = shear; T1 = tension

Rn = 0.6 Fu Anv + Ubs Fu Ant ≤ 0.6 Fy Agv + Ubs Fu Ant (AISC J4-5)

$$\Omega = 2 (ASD)$$

Bolt: M16; nos/side = 3; hole dia. = 0.71 inch; edge dist., e = 1.14 inch

Face Plate

L = 10 inch; W = 5 inch; Th = 5/16 inch

L_1 = 10 - 1.38 = 8.62 inch; $e_1 = k + 0.5 = 0.88 + 0.5 = 1.38$ inch

b = e = 1.14 inch; Th = 5/16 inch

where,

$$Agv = \text{sum of gross area subject to shear yielding} \left(2 \times L1 \times Th\right)$$

$$= 2 \times 8.62 \times 0.313 = 5.39 \, inch^2$$

$$Anv = \text{net area subject to shear rupture} \left(Agv\text{-}\Sigma hole \ area\right)$$

$$= 5.39 - 2 \times 3 \times 0.71 \times 0.313 = 4.06 \, inch^2$$

$$Ant = \text{net area subject to tension} \left(2 \ x \ b \ x \ Th\right)$$
$$= 2 \times 1.14 \times 0.313 = 0.71 \, inch^2$$

Ubs = 1 where tension is uniform; 0.5 where tension is not uniform.

Now,

$$0.6 \, Fu \, Anv + Ubs \, Fu \, Ant = 0.6 \times 65 \times 4.06 + 1 \times 65 \times 0.71 = 208 \, kips$$

$$0.6 \, Fy \, Agv + Ubs \, Fu \, Ant = 0.6 \times 36 \times 5.39 + 1 \times 65 \times 0.71 = 163 \, kips$$

$$Rn = 163 \, kips; V = Rn \, / \, \Omega = 163 \, / \, 2 = 81.5 \, kips$$

Block shear strength, **V = 81.5 kips > R = 20 kips; Safe.**

4.2.5   WEB FRAMING CLEAT ANGLE – IS 800 (LIMIT STATE)

The workout example given below has been done in limit state and using bearing type bolt.

a) Sketch

Refer Figure 4.2

b) **Materials**
   Refer Table 4.6
c) **Member Description**

Supported member: B1	ISMB 250	D = 250 mm
Supporting member: B2	ISMB 250	D = 250 mm

d) **Design Load, R**
   End Shear 70% of web shear capacity. (Assumed for Joint design)
   Web shear:
   Effective depth of web, $D - h1 = 222$ mm; $h1 = 28$ mm
   Web thickness, $t = 6.9$ mm; $h2 = h1 + 15 = 43$ mm
   Shear stress, $\tau sh = (fy/\sqrt{3})/\gamma_{m0} = 250/1.732/1.1 = 131$ MPa
   (IS 800:2007 – 8.4)

$$R = \text{End shear} = 0.70 \times (D - h1) \times t \times \tau sh / 1000$$

$$= 0.7 \times 222 \times 6.9 \times 131 / 1000 = \mathbf{140}\ \text{kN}$$

e) **Cleat sizing**: Provide web cleat angles: **2 ISA 75 75 10**
   Length of cleat, Lv = **175** mm
   Leg thickness, Lt = 10 mm
   Leg size, Lh = 75 mm

   Shear capacity of cleats $= 2 \times 175 \times 10 \times 131 / 1000 = 459$ kN $>$ R; Okay.

f) **Design of Connection Bolts (by limit state)**
   Let us use **M20 HT Bolt 8.8 grade – Bearing type bolt**

   Bolt diameter, d = 20 mm
   $\gamma_{mb} = 1.25$ Material Safety factor.
   Fub = 830 MPa Ultimate tensile stress of bolt.
   Fu = 410 MPa Ultimate tensile stress of plate.
   $n_0 = 0$ Numbers of shear planes with threads intercepting the shear plane.
   $n_s = \mathbf{1}$ Numbers of shear planes without threads intercepting the shear plane.
   Asb = 314 mm^2 Nominal plain shank area of the bolt.
   Anb = 245 mm^2 Net shear area of the bolt at thread.
   $e_1 = 32$ mm End distance along bearing direction.
   $d_0 = 22$ mm Diameter of the hole.
   p = 50 mm Pitch distance along bearing direction (2.5d min).
   kb = 0.485 [kb = smaller of e/3d0, (p/3d0 − 0.25), fub/fu, 1]; Refer Table 4.4.

Shear capacity:

$$Vnsb = (Fub / \sqrt{3}) \times (n_n.Anb + n_s.Asb)(IS\,800:2007 - 10.3.3)$$

$$Vnsb = \left[(830 / 1.732) \times (1 \times 314)\right] / 1000 = 150\,kN$$

Permissible stress in bolt in shear, $fasb = (Vnsb / \gamma_{mb}) / Asb$

$$= (150000 / 1.25) / 314 = 382\,MPa.$$

Bolt value, $Vsb = Asb \times fasb = 314 \times 382 / 1000 = 120\,kN\,(single\ shear)$

Bearing capacity:

Bearing Plate thickness = 6.9 mm [smaller of beam web and cleat thickness].

$Apb = 138\ mm^2$ bearing area on plate ($d \times t$)

$Vnpb$ = nominal bearing strength of *a* bolt = 2.5 kb. *d. t.* Fu (IS 800:2007 – 10.3.4)

$$Vpnb = 2.5 \times 0.485 \times 20 \times 6.9 \times 410 / 1000 = 69\,kN$$

Perm. Bearing stress in bolt / plate, $fapb = (Vnpb / \gamma_{mb}) / Apb$

$$= (69000 / 1.25) / 138 = 400\,MPa$$

Bolt value, $Vsb = Apb.fapb = 138 \times 400 / 1000 = 55\,kN$

Bolt N1 End shear = **140** kN

Number of shear plane, n = 2
Shear Capacity = 2 × 120 = 240 kN; Bearing Capacity = 55kN
Bolt value or Design strength in shear, Vsb = **55** kN per bolt
Number of bolt per side (N1) = 3
Total capacity = 3 × 55 = **165** kN > **R = 140 kN; Safe**.

Bolt N2 End shear = **140** kN

Number of shear plane, n = 1
Shear Capacity = 1× 120 = 120 kN; Bearing Capacity = 55 kN
Bolt value or Design strength in shear, Vsb = **55** kN per bolt.
Number of bolt per side (N2) = 6

$$Total\ capacity = 6 \times 55 = \mathbf{330}kN > R = \mathbf{140}\ kN; Safe.$$

g) Block Shear Capacity, Tb; R = 140 kN

Refer Figure 4.3: Block Shear rupture above

Web Cleat: 2 ISA 75 75 10

$t = 10$ mm; $b_L = bs = 75$ mm; $e_1 = 32$ mm

Length of cleat, Lv = 175 mm

$L = Lv - e_1 = 175 - 32 = 143$ mm; b1 = 32 mm

Connecting bolts: M20 8.8 grade

$d_0 = 22$ mm; nos = 3 per leg

Tdb = Smaller of Tdb1 or Tdb2 (IS 800:2007 – 6.4.1; 6.4.2)

$$Tdb1 = \left[A_{vg}.fy / \left(\sqrt{3}\gamma_{mo} \right) + 0.9\, A_{tn}.fu / \gamma_{m1} \right]$$

Or

$$Tdb2 = \left[0.9 A_{vn}.fu / \left(\sqrt{3}\gamma_{m1} \right) + A_{tg}\, fy / \gamma_{mo} \right]$$

Avg = gross area in shear along bolt line = $2 \times 143 \times 10 = 2860\,\text{mm}^2$

Avn = net area in shear along bolt line = $2860 - 2 \times 3 \times 22 \times 10 = 1540\,\text{mm}^2$

Atg = minimum gross area in tension perpendicular to the line of force

$$= 2 \times 32 \times 10 = 640\,\text{mm}^2$$

Atn = minimum net area in tension perpendicular to the line of force = $530\,\text{mm}^2$

$$Tdb1 = \left[2860 \times 250 / \left(1.732 \times 1.1 \right) + 0.9 \times 530 \times 410 / 1.25 \right] / 1000\,\text{kN} = 532\,\text{kN}$$

$$Tdb2 = \left[0.9 \times 1540 \times 410 / \left(1.732 \times 1.25 \right) + 640 \times 250 / 1.1 \right] / 1000\,\text{kN} = 408\,\text{kN}$$

$$Tdb = 408\,\text{kN}; Tb = Tdb / \gamma_{m1} = 408 / 1.25 = 326\,\text{kN}$$

Block shear capacity, Tb = 326 kN > R = 107 kN; Safe.

4.2.6 WEB FRAMING CLEAT ANGLE – AISC (LRFD)

a) Sketch

Refer Figure 4.2 Shear connection with angle cleats.

b) Material

Refer Table 4.7 Properties of Bolt & Structure – ASTM

c) **Member Descriptions**:

Supported member:	B1	W12 × 26	D = 12.22 inch
Supporting member:	B2	W12 × 26	D = 12.22 inch

d) **Design Load, R**

End Shear 70% of web shear capacity. (Assumed for Joint design)
Effective depth of web, D − h1 = 11.16 inch; h1 = k = 1.063 inch
Web thickness, tw = 0.23 inch; h2 = h1 + 0.75 = 1.813 inch

$$\text{Shear stress}, \tau sh = 0.6\,\text{fy} = 0.6 \times 36 = 21.6\,\text{ksi}$$

$$\text{Web Shear capacity} = \text{Rn} = 0.6\,\text{fy Agv} = 0.6\,\text{fy}\left(D - h1\right).\text{tw}.\phi$$

$$R = \text{Design End shear} = 0.70 \times \left(D - h1\right) \times \text{tw} \times 0.6\,\text{fy} \times \phi, \quad \phi = 0.9$$

$$= 0.7 \times 11.16 \times 0.23 \times 0.6 \times 36 \times 0.9 = \textbf{35 kips} \quad \left(159\text{kN}\right)$$

e) **Cleat Sizing**: Provide web cleat angles: **2 L 3 × 3 × 3/8**

Length of cleat, Lv = 10 inch
Leg thickness, Lt = 3/8 inch (0.375 inch)
Leg size, Lh = 3 inch
Shear capacity of cleats = 2 × 10 × 0.375 × 21.6 × 0.9 = **146** kips > *R*;
  Okay.
ϕ = 0.9 (AISC − G 2.1)

f) **Bolt Sizing**

Bolt: A325X M16 Bearing Type

Bolt dia., d = 16 mm
dh = dia. of hole = 18 mm (AISC − Table J3.3M)
Nominal tensile strength, Fnt = 620 MPa (AISC − Table J3.2)
Nominal shear strength, Fnv = 469 MPa
[X = threads Excluded from shear plane; N = threads Not excluded from
  shear plane.]

Allowable Shear Strengh, Pv = ϕ Rn

Rn = Fnv. Ab (AISC manual J3-1)

$$\text{Fnv} = 469\,\text{MPa}$$

Ab = 201 mm² (Area of bolt in unthreaded body area)

$$\phi = 0.75$$

Bolt value in shear, Pv = 0.75 × 469 × 201/1000 = **71** kN

Bearing strength at Bolt Holes, Pb = φ Rn

$$Rn = 1.2 lctFu \leq 2.4dtFu \, (\text{AISC manual J3} - 6a)$$

where,

　Fu = specified tensile strength = 827 MPa

　d = nominal bolt diameter = 16 mm

　lc = clear distance between edge of hole and the edge of adjacent hole or edge of the material = 20 mm (edge dist. —0.5dh) Refer Table 4.5

　φ = 0.75

　t = thickness of connecting material = 5.84 mm (minimum of web and cleat)

　(= 0.23 inch)

$$1.2 \, lc \, t \, Fu = 1.2 \times 20 \times 5.84 \times 827 / 1000 = 116 \, kN$$

2.4 *d t* Fu = 2.4 × 16 × 5.84 × 827/1000 = 186 kN; **Rn = 116 kN**

Bolt value in bearing, Pb = Rn × φ = 0.75 × 116 = 87 kN

Effective strength of the bolt or Bolt value is the lesser of fastener shear strength or the bearing strength at the bolt hole.

Bolt N1

　End Shear, R = **35** kips

　Number of shear plane, n = 2

　Shear Capacity = 2 × 71 = 142 kN double shear

　Bearing Capacity = 87 kN per bolt

　Bolt value or Design strength in shear, Vsb = **87** kN

　Number of bolts per side (N1) = 3

　Total Capacity = 3 × 87 = 261 kN = **58** kips > *R* = **35** kips; Safe.

Bolt N2

　End Shear, R = 35 kips

　Number of shear plane, n = 1

　Shear Capacity = 1 × 71 = 71 kN; Bearing Capacity = 87 kN

　Bolt value or Design strength in shear, Vsb = 71 kN per bolt.

　Number of bolts per side (N2) = 6

　Total Capacity = 6 × 71 = 426 kN = **94** kips > *R* = **35** kips; Safe.

g) To Determine Block Shear Resistance, $V = \phi\ Rn$

Refer Figure 4.3: Block Shear rupture

$$Rn = 0.6\,Fu\,Anv + Ubs\,Fu\,Ant \le 0.6\,Fy\,Agv + Ubs\,Fu\,Ant\,(AISC\,J4-5)$$

$$\phi = 0.75(LRFD)$$

Web cleat $2L3 \times 3 \times 3/8$

Dimension of a single cleat

e = 1.14 inch; Lv =10 inch; Lt = 3/8 inch; Lh = 3 inch; Cyy = 0.884
Bolt M16 nos/side = 3; hole dia. = 0.71 inch

$$L = 2 \times (10 - 1.14) = 17.72 \text{ inch}$$

$$b = 2 \times (3 - 0.884 = 4.232 \text{ inch}; t = Lt = 3/8 \text{ inch}\,(= 0.375 \text{ in})$$

where,

$$Agv = \text{gross area subject to shear yielding}\,(L \times t\,) = (17.72 \times 0.375) = 6.65 \text{ inch}^2$$

$$Anv = \text{net area subject to shear rupture}\,(Agv\text{-}\ \Sigma\text{hole area})$$

$$= 6.65 - 2 \times 3 \times 0.71 \times 0.375 = 5.05 \text{ inch}^2$$

$$Ant = \text{net area subject to tension}\,(b \times t) = 4.232 \times 0.375 = 1.59 \text{ inch}^2$$

Ubs = 1 where tension is uniform; 0.5 where tension is not uniform.

Now,

$$0.6\,Fu\,Anv + Ubs\,Fu\,Ant = 0.6 \times 65 \times 5.05 + 1 \times 65 \times 1.59 = 300 \text{ kips}$$

$$0.6\,Fy\,Agv + Ubs\,Fu\,Ant = 0.6 \times 36 \times 6.65 + 1 \times 65 \times 1.59 = 247 \text{ kips}$$

$$Rn = 247 \text{ kips}; V = \phi Rn = 0.75 \times 247 = 185 \text{ kips}$$

Block shear strength, V = 185 kips > R = 35 kips; Safe.

It is seen that the number of bolts necessary for end connection of standard sections by LRFD and ASD methods are the same except as noted in following table. Thus, it is suggested that Table 5.1.5 (b) should be used for both methods.

TABLE 4.8
No. of Bolts in AISC – LRFD and ASD

Member Section	LRFD			ASD		
	Bolt Size	Nos N1	Nos N2	Bolt Size	Nos N1	Nos N2
W12 × 26	M16	3	6	M16	4	8
W14 × 30	M16	3	6	M16	4	8
W16 × 36	M16	4	8	M16	5	10
W18 × 40	M16	5	10	M16	6	12
W21 × 50	M20	5	10	M20	6	12
W24 × 84	M24	5	10	M24	6	12
C12 × 20.7	M22	3	6	M22	4	8
C15 × 33.9	M24	4	8	M24	5	10

4.2.7 END CONNECTING FACE PLATE – IS 800 (LIMIT STATE)

a) Sketch

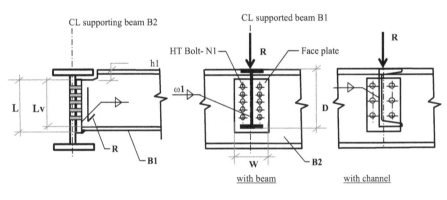

FIGURE 4.6   Shear connection with face plate – bolted.

b) Material
Refer Table 4.6 Properties of Bolt & Structure – IS standard.

c) Member Description

Supported member: B1	ISMB 600	D = 600 mm
Supporting member: B2	ISMB 600	D = 600 mm

d) Design Load, R
End Shear 70% of web shear capacity. (Assumed for Joint design)

Web shear:

Effective depth of web, $D - h1 = 555$ mm; $h1 = 45$ mm

Web thickness, tw = 12 mm; Flange thickness, tf = 20.8 mm

$$\text{Shear stress}, \tau sh = \left(\text{fy} / \sqrt{3}\right) / \gamma_{m0}$$

$$= 250 / 1.732 / 1.1 = 131\,\text{MPa}\left(\text{IS}\,800:2007 - 8.4\right)$$

$$R = \text{End shear} = 0.70 \times \left(D - h1\right) \times \text{tw} \times \tau sh / 1000$$

$$= 0.7 \times 555 \times 12 \times 131 / 1000 = \textbf{611}\,\text{kN}$$

e) Face Plate Sizing

L provided $= D - h1 + 15$ mm $= 570$ mm; Lv $= 555 - 20.8 = 534.2$ mm

W = Width of the plate = width of the flange of the supported beam = 210 mm

Thickness of plate = Thickness of web of supported beam, tp = t = 12 mm

f) Fillet Weld Sizing

Electrode: EX40XX; fu = 410 MPa

Permissible stress in fillet weld, fw = 189 MPa (Shop weld); (fu/ $\sqrt{3}$)/γ_{mw}

$$\omega_1 = \text{fillet weld at shop} = 6\,\text{mm} = \text{thickness of web}$$

Strength of fillet welding connecting supported beam web to face plate,

$$F1 = 2.\text{Lv}.0.7.\omega_1.\text{fw} = 2 \times 534.2 \times 0.7 \times 6 \times 189 / 1000 = 848\,\text{kN} > R;\textbf{Safe}.$$

g) Design of Connection Bolts (by Limit State) (IS 800:2007 – 10.3)

Let us use **M16 HT Bolt 8.8 grade – Bearing type bolt**

Bolt diameter, d = 16 mm

$\gamma_{mb} = 1.25$ Material Safety factor.

Fub = 800 MPa Ultimate tensile stress of bolt.

Fu = 410 MPa Ultimate tensile stress of plate.

$n_0 = 0$ No. of shear planes with threads intercepting the shear plane.

$n_s = \mathbf{1}$ No. of shear planes without threads intercepting the shear plane.

Asb = 201 mm² Nominal plain shank area of the bolt.

Anb = 157 mm² Net shear area of the bolt at thread.

$e_1 = 29$ mm End distance along bearing direction.

$d_0 = 18$ mm Diameter of the hole.

p = 40 mm Pitch distance along bearing direction (2.5d min).

kb = 0.491 [kb = smaller of e/3d0, (p/3d0 − 0.25), fub/fu, 1]; Refer Table 4.4.

Shear Capacity:

$$\text{Vnsb} = \left(\text{Fub} / \sqrt{3}\right) \times \left(n_n.\text{Anb} + n_s.\text{Asb}\right)\left(\text{IS}\,800:2007 - 10.3.3\right)$$

$$\text{Vnsb} = \left[\left(800/1.732\right) \times \left(1 \times 201\right)\right] / 1000 = 93 \text{ kN}$$

Permissible stress in bolt in shear, fasb $= \left(\text{Vnsb} / \gamma_{mb}\right) / \text{Asb}$

$$= \left(93000/1.25\right) / 201 = 370 \text{ MPa}.$$

Bolt value, Vsb $= \text{Asb} \times \text{fasb} = 201 \times 370 / 1000 = 74 \text{ kN} \left(\text{single shear}\right)$

Bearing Capacity:
Bearing plate thickness $= \mathbf{12}$ mm [smaller of beam web and cleat thickness].
    Apb $= 192$ mm^2 bearing area on plate (d $\times$ t)
    Vnpb $=$ nominal bearing strength of *a* bolt $= 2.5$ kb. *d*. *t*. Fu (IS 800:2007 – 10.3.4)

$$\text{Vpnb} = 2.5 \times 0.491 \times 16 \times 12 \times 410 / 1000 = 97 \text{ kN}$$

Perm.Bearing stress in bolt / plate, fapb $= \left(\text{Vnpb} / \gamma_{mb}\right) / \text{Apb}$

$$= \left(97000/1.25\right) / 192 = 404 \text{ MPa}$$

Bolt value, Vsb $= \text{Apb.fapb} = 192 \times 404 / 1000 = 78 \text{ kN}$

Bolt N1 End shear $= \mathbf{611}$ kN
    Number of shear plane, n $= \mathbf{1}$
    Shear Capacity $= 1 \times 74 = 74$ kN; Bearing Capacity $= 78$ kN
    Bolt value or Design strength in shear, Vsb $= \mathbf{74}$ kN per bolt.
    Number of bolt per side (N1) $= 10$
    Total capacity $= 10 \times 74 = \mathbf{740}$ kN $> \mathbf{R = 611}$ kN; **Safe**.
h) **Block Shear Capacity, Tb: R = 611 kN**.

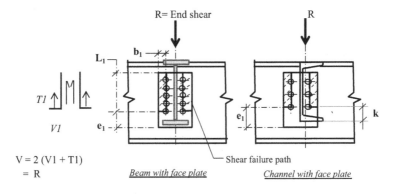

FIGURE 4.7   Block shear rupture.

(V = Block shear strength; V1 = shear; T1 = tension)
   Member section: B1 ISMB 600
   Flange thickness, tf = 20.8 mm
   Face plate: Length, L = 570 mm; Width = W = 210 mm
   Thickness, Th = 12 mm
   Connecting bolts: **M16 8.8 grade**
   D = 16 mm; d_0 = 18 mm, no. of bolts = 5 per side

$$e_0 = 29\,\text{mm}$$

$$e_1 = e_0 + \text{tf} + 15 = 29 + 20.8 + 15 = 64.8\,\text{mm}$$

$$L_1 = L - e_1 = 570 - 64.8 = 505\,\text{mm}; b_1 = 29\,\text{mm}$$

Tdb = Smaller of Tdb1 or Tdb2 (IS 800:2007 – 6.4.1; 6.4.2)

$$\boldsymbol{Tdb1 = \left[A_{vg}.fy / \left(\sqrt{3}\ \gamma_{mo}\right) + 0.9\,A_{tn}.fu\,/\,\gamma_{m1}\right]}$$

Or

$$\boldsymbol{Tdb2 = \left(0.9A_{vn}.fu / \left(\sqrt{3}\gamma_{m1}\right) + A_{tg}\ fy\,/\,\gamma_{mo}\right)}$$

Avg = gross area in shear along bolt line = $2 \times 505 \times 12 = 12120\,\text{mm}^2$

Avn = net area in shear along bolt line = $12120 - 2 \times 5 \times 18 \times 12 = 9960\,\text{mm}^2$

Atg = minimum gross area in tension perpendicular to the line of force

$$= 2 \times 29 \times 12 = 696\,\text{mm}^2$$

Atn = minimum net area in tension perpendicular to the line of force = $588\,\text{mm}^2$

$$Tdb1 = \left[12120 \times 250 / (1.732 \times 1.1) + 0.9 \times 588 \times 410 / 1.25\right] / 1000\,\text{kN} = \mathbf{1764}\ \text{kN}$$

$$Tdb2 = \left[0.9 \times 9960 \times 410 / (1.732 \times 1.25) + 696 \times 250 / 1.1\right] / 1000\,\text{kN} = 1856\,\text{kN}$$

$$Tdb = 1764\,\text{kN}\ Tb = Tdb / \gamma_{m1} = 1764 / 1.25 = 1411\,\text{kN}$$

Block shear capacity, Tb = 1411 kN > R = 611 kN; Safe.

The end connections for all standards sections are designed by the above method. It is found that there are insignificant changes in the number of bolts between the two methods – both working and limit state. Hence, the design chart, Table 5.1.6 (a) may be used for limit state design.

4.2.8   END CONNECTING FACE PLATE – AISC (LRFD)

a) Sketch
Please refer Figure 4.6: Shear connection with face plate – Bolted.

b) Materials
Please refer Table 4.7 Properties of Bolt & Structure – ASTM above.

c) Member Description

Supported member:	B1	W24 × 84	D = 24.1 inch
Supporting member:	B2	W24 × 84	D = 24.1 inch

d) Design Load, R
End Shear 70% of web shear capacity. (Assumed for Joint design)
Effective depth of web, D-h1 = 23.04 inch; h1 = k = 1.06 inch
Web thickness, tw = 0.47 inch; Flange thickness, tf = 0.77 inch
Flange width, bf = 9.02 inch

$$\text{Shear stress}, \tau sh = 0.6\,\text{fy} = 0.6 \times 36 = 21.6\,\text{ksi}$$

$$\text{Web shear capacity} = Rn = 0.6\,\text{fy}\,Agv = 0.6\,\text{fy}\left(D - h1\right).tw.\phi\left(AISC - J4 - 3\right)$$

$$R = \text{Design End shear} = 0.70 \times \left(D - h1\right) \times tw \times 0.6\,\text{fy} \times \phi,\ f = 0.9$$

$$= 0.7 \times 23.04 \times 0.47 \times 0.6 \times 36 \times 0.9 = \textbf{147 kips}\left(667kN\right)$$

e) Face Plate Sizing

$$Lv\ \text{required} = R / \left(tw.0.6\,\text{fy}\right) = 147 / \left(0.47 \times 0.6 \times 36\right) = 14\,\text{inch}.$$

$$L\ \text{provided} = D - h1 + 0.5\,\text{inch}$$

$$L = 24.1 - 1.06 + 0.5 = 24\,\text{inch} > Lv\ \text{reqd.}$$

W = width of the plate = width of the flange of the supported beam
W = 9.02 inch; Provide 10 inch wide plate.

$$\text{Thickness of plate} \geq \text{Thickness of web of supported beam}$$

Tp = tw = 0.47 inch.; Thickness provided = 0.5 inch.

f) Fillet Weld Sizing
Electrode Fe70XXX; fy = 70 ksi; ϕ = 0.75
Permissible Stress in fillet weld, fw = 31.5 ksi (shop weld; fw = 0.6fy. ϕ)

$$\omega_1 = \text{size of fillet weld} = 1 / 4\,\text{inch}\left(0.25\,\text{inch}\right)$$

Strength of fillet welding connecting supported beam web to face plate,

$$F1 = 2.(D - h1 - tf).0.707.\omega_1.fw + (2.bf - tw).0.707.\omega_1.$$

$$= 2(24.1 - 1.06 - 0.77) \times 0.707 \times 0.25 \times 31.5$$
$$+ (2 \times 9.02 - 0.47) \times 0.707 \times 0.25 \times 31.5$$

=246 + 98 = **344** kips > **R = 147 kips**; Safe.

g) **Bolt Sizing**

Bolt: A325X M24; Bearing Type

Bolt dia., d = 24 mm

dh = dia. of hole = 27 mm (AISC – Table J3.3M)

Nominal tensile strength, Fnt = 620 MPa (AISC – Table J3.2)

Nominal shear strength, Fnv = 469 MPa

[X = threads Excluded from shear plane; N = threads Not excluded from shear plane.]

Allowable Shear Strengh, Pv = ϕ Rn

$$Rn = Fnv.Ab (AISC \text{ manual } J3 - 1)$$

$$Fnv = 469 \, MPa$$

$$Ab = 453 \, mm^2 (\text{area of bolt in unthreaded body area})$$

$$\phi = 0.75$$

Bolt value in shear, $Pv = 0.75 \times 469 \times 453 / 1000 = $ **159** kN

Bearing strength at Bolt Holes, Pb = ϕ Rn

$$Rn = 1.2 \, lc \, t \, Fu \leq 2.4 \, d \, t \, Fu (AISC \text{ manual } J3 - 6a)$$

where,

Fu = specified tensile strength = 827 MPa

d = nominal bolt diameter = 24 mm

lc = clear distance between edge of hole and the edge of adjacent hole or edge of the material = 30.5 mm; (edge dist. – 0.5dh) Refer Chapter 4 Table 4.5

$\phi = 0.75$

t = thickness of connecting material = 12 mm (minimum of web and cleat)

(= 0.472 inch)

$$1.2 \, lc \, t \, Fu = 1.2 \times 30.5 \times 12 \times 827 \, / \, 1000 = 363 \, kN$$

$$2.4 \, d \, t \, Fu = 2.4 \times 24 \times 12 \times 827 \, / \, 1000 = 572 \, kN; \mathbf{Rn = 363kN}$$

Bolt value in bearing, $Pb = Rn \times \phi = 0.75 \times 363 = \mathbf{272}$ kN

Effective strength of the bolt or Bolt value is the lesser of fastener shear strength or the bearing strength at the bolt hole.

Bolt N1
End Shear, R = **147** kips
Number of shear plane, n = **1**
Shear Capacity = 1 × 159 = 159 kN single shear
Bearing Capacity = 272 kN per bolt
Bolt value or Design strength in shear, Vsb = **159** kN
Number of bolts per side (N1) = **10**
Total Capacity = 10 × 159 = 1590 kN = **351** kips > *R* = **147** kips; Safe.

h) Block Shear Resistance, V = ϕ Rn; R = 147 kips
Refer Figure 4.7: Block Shear rupture
(V = Block shear strength; V1 = shear; T1 = tension)

$$Rn = 0.6 \, Fu \, Anv + Ubs \, Fu \, Ant \le 0.6 \, Fy \, Agv + Ubs \, Fu \, Ant \left(AISC \, J4 - 5 \right)$$

$$\phi = 0.75 \left(LRFD \right)$$

Member Section: W24 × 84
Face plate:
    L = 24 inch; W = 10 inch; Th = 1/2 inch
Connecting bolt:
    Bolt – M24; no. of bolts = 10 (5 per side); hole dia. = 1.06 inch
    e = 1.73 inch (edge distance)

$$L_1 = 24 - 1.56 = 22.44 \, inch; e_1 = k + 0.5 = 1.56 \, inch$$

$$b = e = 1.73 \, inch; Thickness, Th = 1 \, / \, 2 \, inch$$

where,

Agv = sum of gross area subject to shear yielding $\left(2 \times L1 \times Th \right)$

$$= 2 \times 22.41 \times 0.5 = 22.44 \, inch^2$$

Anv = net area subject to shear rupture $\left(Agv- \, Shole \, area \right)$

$$= 22.44 - 2 \times 5 \times 1.06 \times 0.5 = 17.14 \, \text{inch}^2$$

Ant = net area subject to tension $(2 \times b \times Th) = 2 \times 1.73 \times 0.5 = 1.73 \, \text{inch}^2$

UBS = 1, where tension stress is uniform; 0.5 where tension is non-uniform
Now,

$$0.6 \, Fu \, Anv + Ubs \, Fu \, Ant = 0.6 \times 70 \times 17.14 + 1 \times 70 \times 1.73 = 841 \, \text{kips}$$

$$0.6 \, Fy \, Agv + Ubs \, Fu \, Ant = 0.6 \times 36 \times 22.44 + 1 \times 70 \times 1.73 = 606 \, \text{kips}$$

Rn = 606 kips; $V = \phi$ Rn = 0.75 × 606 = 455 kips
  Block shear strength, **V = 455** kips > **R = 147 kips**; Safe.
  There are minor changes in the number of bolts required for end connection with face plate while calculating for LRFD and ASD methods. Thus, it is suggested to use the design chart Table 5.1.6 (b) provided in Chapter 5 for both LRFD and ASD load combinations.

4.3   BEAM TO COLUMN – MOMENT CONNECTION WITH BEARING TYPE BOLT

a) **Reference**
  1. AISC Steel construction manual – 14 Edition
  2. IS 800:2007 / IS 816:1969 / IS 1367:1967
b) **Material**

TABLE 4.9
Properties of Structural Materials for MC Joint

Yield stress	f_y	36	ksi	250	MPa
Tensile stress	F_u	65	ksi	450	MPa
Plate bending stresses	Tension	σbt		150	MPa
	Compression	σbc		165	MPa
Weld electrode FE70XXX		70	ksi	483	MPa
	fw	42	ksi	290	MPa
HT Bolt 8.8 ($d <= 16$)	Fyb	640	Fub	800	MPa
HT Bolt 8.8 ($d > 16$)	Fyb	660	Fub	830	MPa
Elasticity	E	29000	ksi	200000	MPa

---

c) Sketch

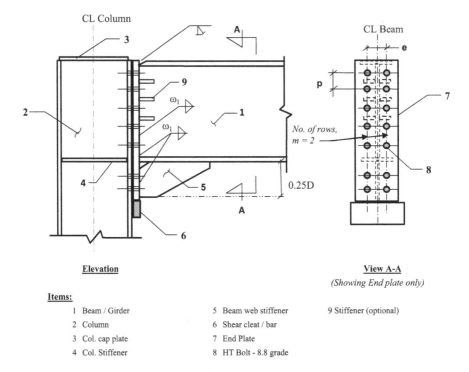

Items:

1	Beam / Girder	5 Beam web stiffener	9 Stiffener (optional)
2	Column	6 Shear cleat / bar	
3	Col. cap plate	7 End Plate	
4	Col. Stiffener	8 HT Bolt - 8.8 grade	

FIGURE 4.8   Moment connection joint – with bearing type HT bolt.

d) Design Parameters

Design forces:

Beam End Moment, M = **5440** kNm; Beam End Shear, **V = 1722** kN

Beam/Girder section:

Flange plates – 600 mm × 32 mm; Web plate – 1536 mm × 16 mm

Web stiffener – 385 mm × 16 mm

Column section:

Flange plates – 600 mm × 28 mm; Web plate – 1144 mm × 28 mm

End plate:

Height h (=D) = 1985 mm; Wide, be (=B) = 625 mm

Thick, Th = 32 mm; e = 125 mm

(e = 3 × bolt dia. + web thk including weld size + clearance)

HT Bolt 8.8 grade:

Dia. ϕ = 27 mm; No. of vertical rows, m = 2; Nos per row = 25 (2.5ϕ)

Total no. of bolts, N = 50; p = 79 mm

Unsupported length:

Beam/girder – Span, Lx = 12 m; Lateral restraint, Ly = 2.5 m

Column – Height, Lx = 12 m; Lateral support, Ly = 6 m

e) To determine Design Strength of Bolt – Bolt Value (According to IS 800 – 2007)

Shear Capacity

$$Vdsb = Vnsb / \gamma mb$$

where,

    Vnsb = Nominal shear capacity of bolt.

    γ_{mb} = Partial safety factor of bearing type bolt

    γ_{mb} = 1.25 (Table 5 – IS 800:2007)

    $Vnsb = (Fub / \sqrt{3}) \times (n_n.Anb + n_s.Asb)$

where,

    Fub = Ultimate tensile stress of bolt.

    n_n = No. of shear planes with threads intercepting the shear plane.

    ns = No. of shear planes without threads intercepting the shear plane.

    Asb = Nominal plain shank area of the bolt.

    Anb = Net shear area of the bolt at thread (area corresponding to root dia. at thread).

Shear capacity of a bolt:

$$\text{Bolt diameter, d} = 27\,mm$$

$$\gamma_{mb} = 1.25$$

$$Fub = 830\,MPa$$

$$n_n = 0$$

$$n_s = 1$$

$$Asb = 573\,mm^2$$

$$Anb = 459\,mm^2$$

$$Vnsb = (830 / 1.732) \times (1 \times 573) / 1000 = 275\,kN$$

$$Vdsb = 275 / 1.25 = 220\,kN.$$

Bearing Capacity

$$Vdpb = Vnpb / \gamma_{mb}$$

Vnpb = nominal bearing strength of a bolt = 2.5 kb.d.t.Fu

where,

e1 = End distance along bearing direction.

p = Pitch distance along bearing direction.

d_0 = Diameter of the hole.

Fub = Ultimate tensile stress of the bolt.

Fu = Ultimate tensile stress of plate.

d = Nominal diameter of the bolt.

t = Summation of thicknesses of the connected plate experiencing bearing stress.

Now,

Bolt dia., d = 27 mm; e_1 = 50 mm (1.5 × d or 50 mm, which one is greater.)

d_0 = 30 mm; P = 81 mm (3 × d); Fub = 830 MPa

Fu = 450 MPa; t = 28 mm (smaller of Beam end plate and Column flange thickness)

γ_{mb} = 1.25; kb = 0.556

Bearing area, Apb = 840 mm^2 ($d \times t$)

$$\text{Vpnb} = 2.5 \times 0.556 \times 27 \times 28 \times 450 / 1000 = 473 \, \text{mm}$$

$$\text{Vdpb} = 473 / 1.25 = 378 \, \text{kN}$$

Vdb = smaller of Vdsb and Vdpb; **Vdb = 220 kN**.

Tension Capacity: [IS 800:2007 – CL 10.3.5]

$$\text{Tb} \leq \text{Tdb}$$

where

Td = factored tensile force

$$\text{Td} = \text{Tnb} / \gamma mb$$

Tnb = 0.9. Fub. An < Fyb. Asb. $(\gamma_{mb}/\gamma_{m0})$ [Nominal Tension capacity.]

Bolt dia., d = 27 mm; Fub = 830 MPa

Fyb = 660 MPa; γ_{mb} = 1.25; γ_{m0} = 1.1

An = 459 mm^2 [net tensile stress area]

Asb = 573 mm^2 [shank area of bolt]

$$0.9.\text{Fub.An} = 0.9 \times 830 \times 459 / 1000 = 343 \, \text{kN}$$

$$\text{Fyb.Asb.}(\gamma_{mb} / \gamma_{m0}) = 660 \times 573 \times (1.25 / 1.1) / 1000 = 430 \, \text{kN}$$

Tnb = 343 kN

By working stress

Shear capacity

Permissible stress in bolt in shear, fasb = 0.6 Vnsb/Asb = (0.6 × 275/573) × 1000 = 288 MPa

Bolt value, Vsb = Asb *x* fasb = 573 × 288/1000 = 165 kN

Bearing capacity

Perm. Bearing stress in bolt/plate, fapb = 0.6 Vnpb/Apb = (0.6 × 473/840) × 1000 = 338 MPa

Bolt value, Vsb = Apb. fapb = 840 × 338/1000 = 284 kN

Bolt value or Design strength, **Vsb = 165 kN**

Tension capacity

Permissible tensile stress of the bolt, fatb = 0.6 Tnb/An = (0.6 × 343/459) × 1000 = 448 kN

$$\text{Bolt value in tension} = \text{An.fatb} = 459 \times 448 / 1000 = \mathbf{206 \ kN}$$

Bolt value or Design tension capacity, Td = 206 kN
   Fillet weld sizing
   End Shear, V = 1722 kN = 380 kips
   AISC code
   Welding Electrode Fe70XXX; fy = 70 ksi; $\Omega = 2$
   Permisible Stress in fillet weld, fw = 21 ksi (shop weld fw = 0.6fy/Ω)

$$\omega_1 = \text{fillet weld size} = 1 / 4 \text{ inch} = 6 \text{ mm}$$

Strength of fillet welding connecting supported beam web to end plate,

$$F1 = 2 \left(\text{web} + \text{stiffener} \right) 0.707 \ \omega_1 \ \text{fw} = 2 \times \left(60 + 15 \right) \times 0.707 \times 0.25 \times 21$$

=**561 kips** > *V*; Safe.

IS code
   Electrode: EX40XX; fy 330 MPa
   Permissible Stress in fillet weld, fw = 132 MPa [shop weld fw = 0.4fy (IS 800:2007 11.6.3)]
   ω_1 = fillet weld size = 6 mm
Strength of fillet welding connecting supported beam web to end plate,

$$F1 = 2 \, \text{Lv}.0.7.\omega_1.\text{fw} = 2 \times \left(1536 + 385 \right) \times 0.707 \times 6 \times 132 / 1000$$

$$= \mathbf{2130 \ kN} > \text{V; Safe.}$$

f) To Determine NA Depth and Stresses in Bolt

Yield stress of steel plate, fy = 250 N/mm^2

Bending stress of plate, σ_{bt} = 150 N/mm^2

$$\text{Net Cross - sec. area of bolt} = \text{A mm}^2$$

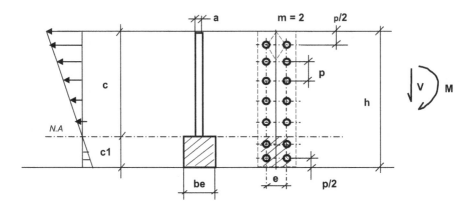

FIGURE 4.9   Stress in bolts.

No. of rows = m
Vertical spacing of bolts = p mm
Bolt center distance = e mm

Now, let us consider

$$p = h / n$$

$$a = (A / p).m$$

$$1/2.a.c^2 = 1/2.be.c_1^2$$

$$c_1 / c = \sqrt{(a / be)}$$

$$I = (a.c^3 / 3) + (be.c_1^3 / 3)$$

$\sigma t_{cal} = M. (c - p/2)/I$; Max Tension / bolt, $T = \sigma t_{cal} \times A$

$$c = h - c_1$$

$c = h - c x$, where $x = c_1/c$

$$c(1 + x) = h$$

$$c = h / (1 + x)$$

$$c_1 + c = h$$

End Plate: Depth, D = h and Width, B = be
Now, putting the values
e = 125 mm (e = 3 bolt dia.+ web thk. including weld size + clearance)

$$A = 573 \, \text{mm}^2$$

$$p = 1985 / 25 = 79 \, \text{mm}$$

$$a = (573 / 79.4) \times 2 = 14.43 \, \text{mm}$$

$$x = c_1 / c = \sqrt{(a / be)} = \sqrt{(14.43 / 625)} = 0.15$$

$$c = h / (1 + x) = 1985 / (1 + 0.15) = 1726 \, \text{mm}$$

$$c_1 = h - c = 1985 - 726 = 259 \, \text{mm} \, (\text{NA at } 1/7 \text{ of depth above bottom})$$

$$I = (a.c^3 / 3) + (be.c_1^3 / 3) = 28352046655 \, \text{mm}^4$$

$$M = 5440 \, \text{kNm} = 5440000000 \, \text{Nmm}$$

$$\sigma t_{cal} = M.(c - p/2) / I = 5440000000 \times (1726 - 0.5 \times 79.4) / 28352046655$$

$$= 324 \, \text{MPa}$$

g) **To Determine Forces in Bolt**
Max Tension /bolt, $T = \sigma t_{cal} \times An = 324 \times 459/1000 = 149 \, \text{kN}$
Allowable strength, Td = 206 kN > T; Safe.
Max shear/bolt, Vsh = V/N = 1722/50 = 34 kN
Allowable strength, Vsb = 165 kN > Vsh; Safe.
Stress interaction ratio subjected to combined tension and shear (working stress):
   149/206 + 34/165 = 0.72 + 0.21 = 0.93 Safe < 1.

h) **End Plate**
T = 149 kN
e = 125 mm
p = 79 mm
Th = 32 mm end plate thickness

$$M = 149000 \times 125 / 4$$
$$= 4656250 \, \text{Nmm}$$

Effective width = 158 mm
   (with Stiffener mkd. 9)

$$fbc = M / Z = 4656250 / (158 \times 32^2 / 6) = \mathbf{173} \, \text{MPa}$$

Allowable bending stress = 0.75 fy = **188** MPa > fbc; Safe.
[CL 11.3 3 IS 800:2007]

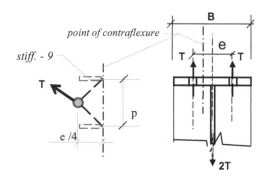

FIGURE 4.10 End plate.

i) Summary of Calculation Results

Table 5.2.2 in Chapter 5 provides details of connections materials for Full strength Bolted joint of Standard sections.

TABLE 4.10
Summary of Calculation – MC Joint Bolted

Beam	2 Flng	600	mm	32	mm
	1 Web	1536	mm	16	mm
	Stiff	385	mm sq.	16	mm
Column	2 Flng	600	mm	28	mm
	1 Web	1144	mm	28	mm

Connection Materials:

End Plate

height		wide		thick	
1985	mm	625	mm	32	mm

Bolt HT 8.8 Grade; Bearing type

	Nos	Dia.mm	Rows	Spcg. in mm	
				Vert	Horz
	50	27	2	79	125

Stress Interaction Ratio = 0.93 < 1; Safe.

4.4 TRUSS MEMBER JOINTS

4.4.1 DESCRIPTION

Truss members are joined by a common gusset plate. The members are connected to the gusset plate by bolts or weld for transferring the forces in member (tension or compression force). The CG lines of each member meet at a point. The shape of the gusset is developed matching the array of bolts as necessary for transferring forces at each member's end.

The member's forces are usually given in the design drawing for joint design; otherwise, the designer should use full strength of the member (tension or compression) to determine the number of bolts required for connection.

In this section, we will provide a way to calculate the full strength of a member section, strength of bolts, and numbers required for full strength in jointing.

4.4.2   SINGLE ANGLE SECTIONS – AISC (ASD)

a) **Material**

Please refer Table 4.7.

b) **Design Parameters**

Size of angle: **L 4 × 4 × 1/2**

Bolt: M24 A325X

Nominal dia., d = 24 (1 inch)

No. of bolts = 4

Dia. of bolt hole, do = 27 mm (1.06 inch)

Gusset thickness, t = 12 mm (1/2 inch)

Pitch distance, p = 72 mm (2-5/6 inch)

Edge distance, e = 44 mm (1-3/4 inch)

b_L = bs = 4 inch; t = 0.5 inch

Ag = 3.75 inch²; An = 3.22 inch²

w = 4 inch = 102 mm; w1 = 1.18 inch = 30 mm

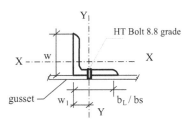

FIGURE 4.11   Single angle bolted joint – section.

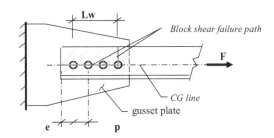

FIGURE 4.12   Single angle bolted joint – elevation.

c) **To Determine Tensile strength, T**

Tensile strength, T shall be the lower value of tensile yielding in gross section (Pn1) and tensile rapture in net section (Pn2).

Tensile yielding, Pn1 = FyAg/Ω = 36 × 3.75/1.67 = 80.84 kips

ϕ = 0.9 LRFD; Ω = 1.67 ASD

Tensile rupture, Pn2 = Fu Ae/Ω = Fu An. U/Ω = 65 × 3.22 × 0.86/2 = 90 kips

ϕ = 0.75 LRFD; Ω = 2 ASD

Ae = An.U, U = shear lag factor as in Table D3.1 (AISC manual)

$$U = 1 - x/L; L = \left[(4-1) \times 72\right] = 216 \, mm = 8.5 \, inch$$

where,

L = Length of connection (distance between end bolts)

x = Cyy = 30 mm; $U = 1 - (36/216) = 0.861$

U = 0.8 with 4 fastener (Case 8) and 0.861 with case 2

Allowable Tension, T = 80.84 kips

d) **To Determine Compressive strength, C**

Member size: **L 4 × 4 × 1/2**

Member slenderness check: $b/t < = 0.45 \sqrt{(E/Fy)}$

$$0.45 \sqrt{(E/Fy)} = \sqrt{0.45(29000/36)} = 12.8$$

$b/t = 4/0.5 = 8$ (Non Slender)

Allowable Compressive strength, **C** = Pn3/Ω = 116/1.67 = **69** kips

$$\left[\Omega = 1.67 \, (ASD)\right]$$

Nominal compressive strength, Pn = Fcr. Ag = 31 × 3.75 = 116 kips (AISC E3-1)

Length of member, Le = 48 inch (Assumed)

K = 0.85; r_{min} = 0.776

$$\lambda = KL/r = 0.85 \times 48/0.776 = 53$$

$$4.71\sqrt{(E/Fy)} = 4.71\sqrt{(29000/36)} = 134$$

Elastic buckling stress, Fe = $\pi^2 E/l2$ = 3. 14^2 × 29000/53^2 = 102 ksi

KL/r < = 4.71 $\sqrt{(E/Fy)}$. Hence, Fcr = [0. 658$^{(Fy/Fe)}$] × Fy

$$Fcr = \left[0.658^{(36/102)}\right] \times 36 = 31 \, ksi$$

Allowable comp., C = 69 kips

e) To Determine Block Shear Resistance, $V = Rn/\Omega$

$$Rn = 0.6\,Fu\,Anv + Ubs\,Fu\,Ant \leq 0.6\,Fy\,Agv + Ubs\,Fu\,Ant\,(AISC\,J4-5)$$

$$\phi = 0.75\,(LRFD); \Omega = 2\,ASD$$

where,

$$Agv = \text{gross area subject to shear yielding} \left[(L+e).t\right]$$
$$= (8.5+1.73)\times0.5 = 5.12\,inch^2$$

$$Anv = \text{net area subject to shear rupture}\,(Agv\text{-}\Sigma hole\ area)$$

$$= 5.12 - 4\times1.06\times0.5 = 3\,inch^2$$

$$Ant = \text{net area subject to tension}\,(b.t) = (4-1.18)\times0.5 = 1.41\,inch^2$$

Ubs = 1, where tension stress is uniform (0.5 where tension stress is non-uniform)

Now,

$$0.6\,Fu\,Anv + Ubs\,Fu\,Ant = 0.6\times65\times3+1\times65\times1.41 = 209\,kips$$

$$0.6\,Fy\,Agv + Ubs\,Fu\,Ant = 0.6\times36\times5.12 +1\times65\times1.41 = 202\,kips$$

Rn = 202 kips; $V = Rn/\Omega = 202/2 = 101$ kips
Block shear strength, V = 101 kips > T; Okay.

Let us consider the maximum of tensile and compressive strength for a full strength joint design. (Design value of F should be minimum of tension and compression force if actual length of member is known.)

Full strength, F = 80.84 kips (367 kN)

f) Bolt Sizing

Let us use: **Bolt: M24; A325X**

d, bolt dia. = 24 mm; d_0 = dia. of hole = 27 mm;(AISC – Table J3.3M)

Nominal tensile strength, Fnt = 620 MPa

Nominal shear strength, Fnv = 469 MPa

[X = threads Excluded from shear plane; N = threads Not excluded from shear plane.]

Allowable shear strengh, Pv = Rn/Ω

$$Rn = Fnv.Ab \qquad (AISC\ manual\ J3-1)$$

Bolt value in Shear, Pv = 469 × 453/2/1000 = **106 kN = 23 kips**

Bearing strength at Bolt Holes, Pb = Rn/Ω

$$Rn = 1.2 \, lc \, t \, Fu \leq 2.4 \, d \, t \, Fu \, (\text{AISC manual J3} - 6a)$$

$$Fnv = 469 \text{ MPa} \quad Ab = 453 \text{ mm}^2 \; (\text{area of bolt in unthreaded body area})$$

$$\Omega = 2$$

where,

Fu = specified tensile strength = 450 MPa

d = nominal bolt diameter = 24 mm

lc = clear distance between edge of hole and the edge of adjacent hole or edge of the material = **44** mm [Min of (3d -d) or as per chart (US Bolt)]

$\Omega = 2$

t = thickness of connecting material = 12 mm (0.5 inch) (minimum of leg thickness and gusset plate)

$$1.2 \, lc \, t \, Fu = 1.2 \times 44 \times 12 \times 450 \, / \, 1000 = 285 \text{ kN}$$

$$2.4 \, d \, t \, Fu = 2.4 \times 24 \times 12 \times 450 \, / \, 1000 = 311 \text{ kN}$$

$$Rn = 285 \text{ kN}$$

Bolt value in Bearing, Pb = 285/2 = **143**kN **31 kips**

Effective strength of the bolt or Bolt value is the lesser of fastener shear strength or the bearing strength at the bolt hole.

So, the **Bolt value, P = 106 kN = 23 kips**

Bolt capacity; Design Load = **80.84** kips

Number of shear plane, n = **1**

Shear Capacity = 1 × 106 = 106 kN

Bearing Capacity = 143 kN

Bolt value or Design strength in shear, Vs = **106** kN per bolt

Number of bolts = 4

Total Capacity = 4 × 106 = 424 kN = **93 kips > F; 80.84 kips; Safe**.

4.4.3   Double Angle Sections – AISC (ASD)

The double angle sections are commonly used in various types of roof trusses. The angles are placed back to back and separated by gussets at joints and pack plates along the length of members at specified intervals. The design of joints with double angles are given in the workout example below.

a) Material

Please refer Table 4.7.

b) **Design Parameters**

Size of angle: **2 – L 3 ×3 × 5/16**

Bolt: **M20**; A325X

Nominal dia., d = 20 mm (0.79 inch)

No. of bolts = 4

Dia. of bolt hole, do = 22 mm (0.87 inch)

Gusset thickness, t = 12 mm (1/2 inch)

Pitch distance, p =60 mm (2-1/3 inch)

Edge distance, e = 32 mm (1-1/4 inch)

[X = threads Excluded from shear plane; N = threads Not excluded from shear plane.]

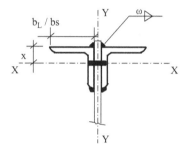

FIGURE 4.13   Double angle bolted joint – section.

Properties of double angle member:

b_L = bs = 3 inch; t = 0.31 inch; Ag = 3.56 inch^2; An = 3.02 inch^2

w = 3 inch = 76 mm; w_1 = 0.86 inch = 22 mm

wt = 12.2 lb/ft; r_{min} = r = 0.918 inch

L = 4 ft (assumed for compression strength determination)

$$b = b_L - Cyy = 3 - 0.86 = 2.14 \text{ inch}$$

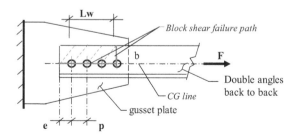

FIGURE 4.14   Single angle bolted joint – elevation.

Length, Lw = 180 mm = **7.09** inch

c) Member Force

To determine Tensile strength, T:

Tensile strength, T shall be the lower value of tensile yielding in gross section (Pn1) and tensile rapture in net section (Pn2).

Tensile yielding, $Pn1 = FyAg/W = 36 \times 3.56/1.67 = 77$ kips (AISC – D2-1)

$\phi = 0.9$ LRFD; $\Omega = 1.67$ ASD

Tensile rupture, $Pn2 = Fu\,Ae/\Omega = Fu\,An.\,U/\Omega = 65 \times 3.02 \times 0.88/2 = 86$ kips

$\phi = 0.75$ LRFD; $\Omega = 2$ ASD

$Ae = An.U$, U = shear lag factor as in Table D3.1 (AISC manual)

$$U = 1 - x/L; L = Lw = 180\,mm = 7.09\,inch$$

where,

L = Length of connection (distance between outermost bolts)

$x = Cyy = 22$ mm; $U = 1 - (22/180) = 0.88$

$U = 0.8$ with 4 fastener (Case 8) and 0.88 with case 2

Allowable Tension, T = 77 kips

To determine Compressive strength, C:

Member size: **L 3 × 3 × 5/16**

Effective length, $Lc = 48$ inch (assumed)

$$\text{Slenderness ratio} = Lc/r = 48/0.918 = 52$$

Pn = Fcr.Ag

where,

Fcr = Critical stress and Ag = gross c/s area. (AISC 360-16 Eqn E3-1)

When, $Lc/r < = 4.71\sqrt{(E/Fy)} = 4.71\sqrt{(29000/36)} = 134$ (AISC 360-16 Eqn E3-2)

$$Fcr = \left[0.658^{(Fy/Fe)}\right].\,Fy = \left[0.658^{(36/105)}\right] \times 36 = \mathbf{31}\ ksi$$

and if,

$Lc/r > 4.71\,(E/Fy)\,^\wedge\,0.5 = 134$ (AISC 360-16 Eqn E3-3)

$$Fcr = 0.877\,Fe = 0.877 \times 105 = \mathbf{92}ksi$$

$$Fe = \text{elastic buckling stress} = \pi^2 E/\left(Lc/r\right)^2 = \left(3.14^2 \times 29000\right)/52^2 = 105\ ksi$$

$E = 29000$ ksi; $Fy = 36$ ksi; $Ag = 4$ inch^2

$$Lc/r = 52\ < 134$$

Hence, **Fcr = 31 ksi**

Pn = Fcr. Ag = 31 × 3.56 = **111** kips (504 kN)
Pr = Pn/Ω = 111/1.67 = **66** ksi (301 kN)
Allowable comp., C = 66 kips
To determine block shear resistance, V = Rn/Ω:

$$\text{Rn} = 0.6 \, \text{Fu Anv} + \text{Ubs Fu Ant} \leq 0.6 \, \text{Fy Agv} + \text{Ubs Fu Ant} \left(\text{AISC J4} - 5 \right)$$

$$\phi = 0.75 \left(\text{LRFD} \right); \Omega = 2 \, \text{ASD}$$

where,

$$\text{Agv} = \text{gross area subject to shear yielding} \left[\left(\text{Lw} + e \right).t \right] = 2 \left(7.09 + 1.26 \right) \times 0.31$$

$$= 5.18 \, \text{inch}^2$$

Anv = net area subject to shear rupture (Agv-Σ hole area)

$$= 5.18 - 4 \times 0.87 \times 2 \times 0.31 = 3.02 \, \text{inch}^2$$

Ant = net area subject to tension $\left(2 \, \text{b.t} \right) = 2 \times 2.14 \times 0.31 = 1.33 \, \text{inch}^2$

Ubs = 1, where tension stress is uniform (0.5 where tension stress is non-uniform)

Now,

$$0.6 \, \text{Fu Anv} + \text{Ubs Fu Ant} = 0.6 \times 65 \times 3.02 + 1 \times 65 \times 1.33 = 204 \, \text{kips}$$

$$0.6 \, \text{Fy Agv} + \text{Ubs Fu Ant} = 0.6 \times 36 \times 5.18 + 1 \times 65 \times 1.33 = 198 \, \text{kips}$$

Rn = 198 kips V = Rn/Ω = 198/2 = 99 kips
Block shear strength, V = 99 kips > T; Okay.
Let us consider the maximum of tensile and compressive strength for a full strength joint design.
Full strength, F = 77 kips
d) **Bolt Sizing**
Let us use **Bolt: M20; A325X**
d, bolt dia. = 20 mm; d_0 = dia of hole = 22 mm (AISC – Table J3.3M)
Nominal tensile strength, Fnt = 620 MPa (AISC – Table J3.2)
Nominal shear strength, Fnv = 469 MPa
[X = threads Excluded from shear plane; N = threads Not excluded from shear plane.]
Allowable shear strengh, Pv = Rn/Ω
Rn = Fnv. Ab (AISC manual J3-1)
Fnv = 469 MPa; Ab = 314 mm^2 (area of bolt in unthreaded body area)

$$\Omega = 2$$

Bolt value in shear, $Pv = 469 \times 314/2/1000 = 74$ kN = **16.64 kips**
Bearing strength at Bolt Holes, $Pb = Rn/\Omega$

$$Rn = 1.2lctFu \leq 2.4dtFu \left(\text{AISC manual J3} - 6a\right)$$

where,

Fu = specified tensile strength = 450 MPa

d = nominal bolt diameter = **20** mm

lc = clear distance between edge of hole and the edge of adjacent hole or edge of the material = **32** mm [Min of (3d -d) or as per chart (US Bolt)]

$\Omega = 2$

t = thickness of connecting material = 12 mm (0.5 inch) (minimum of leg thickness and gusset plate)

$$1.2 \, lc \, t \, Fu = 1.2 \times 32 \times 12 \times 450 / 1000 = 207 \, kN$$

$$2.4 \, d \, t \, Fu = 2.4 \times 20 \times 12 \times 450 / 1000 = 259 \, kN$$

$Rn = 207$ kN

Bolt value in Bearing, $Pb = 207/2 = 104$ kN = **23.27 kips**
Bolt capacity Design Load = **77** kips
Number of shear plane, n = **2**
Shear Capacity = $2 \times 16.64 = 33.28$ kips
Bearing Capacity = 23.27 kips
Bolt value or Design strength in shear, $Vs = $ **23** kips per bolt
Number of bolts = 4
Total Capacity = $4 \times 23 = $ **92 kips > F**; **77 kips**; **Safe**.

4.4.4   SINGLE ANGLE SECTIONS – IS 800 (WORKING STRESS)

a) **Material Properties**
   Refer Table 4.6
b) **Design Parameters**
   Refer Figures 4.11 and 4.12 for Sectional view and Elevation.
   Size of angle: **ISA 100 × 100 × 10**
   Bolt: **M24** High tensile 8.8 grade
   Nominal dia., d = 24 mm; No. of bolts = 4
   Dia. of bolt hole, do = 27 mm; Gusset thickness, tg = 12 mm
   Pitch distance, p = 72 mm (3 × d)
   Edge distance, e = 44 mm

b_L = bs = 100mm; t = 10 mm; Ag = 1903 mm²; An = 1633 mm²

w = 100 mm; w_1 = 28.4 mm

c) **To Determine Tensile Strength, Ts:**

Tensile strength, Ts shall be the lower value of tensile yielding in gross section (Ts1) and tensile rapture in net section (Ts2).

1. Tensile yielding, Ts1 = Ag (0.6fy) = 1903 × (0.6 × 250)/1000 = **285** kN
   (IS 800:2007 – 11.2.1)
2. Tensile rupture, Ts2 = 0.69 Tdn (IS 800:2007 – 11.2.1)

$$\text{Tdn} = 0.9 \, \text{Anc fu} / \gamma_{m1} + \beta \, \text{Ago fy} / \gamma_{mo} \left(\text{IS} \, 800 : 2007 - 6.3.3 \right)$$

$$\beta = 1.4 - 0.076 \left(w / t \right) \left(fy / fu \right) \left(bs / Lc \right) <= 0.9 \, fu \gamma_{mo} / fy \gamma_{m1} >= 0.7$$

where,

w = outstanding leg width = 100 mm (IS 800:2007 – 6.3.4)

bs = shear lag width = w + wl = 100 + 28.4 = 128.4 mm

Lc = length of end connection = distance between two outermost bolt

$$= \left(4 - 1 \right) \times 72 = 216 \, \text{mm}$$

Anc = net area of the connected leg = 730 mm²

Ago = gross area of outstanding leg = 1000mm²

t = thickness of the leg = 10 mm

β = lower of the following values of (i) and (ii) but more than or equal to 0.7

(i) $1.4 - 0.076 \left(w / t \right) \left(fy / fu \right) \left(bs / Lc \right)$
      $= 1.4 - 0.076 \left(100 / 10 \right) \times \left(250 / 410 \right) \times \left(128.4 / 216 \right) = 1.125$

(ii) $0.9 \, fu \, \gamma_{mo} / fy \, \gamma_{m1} = 0.9 \times 410 \times 1.1 / \left(250 \times 1.25 \right) = 1.299$

(iii) ≥ 0.7

So, **β = 1.125**

Now,

$$\text{Tdn} = 0.9 \times 730 \times 410 / 1.25 + 1.125 \times 1000 \times 250 / 1.1 = 471178 \, \text{N}$$

$$= 471 \, \text{kN}$$

$$Ts2 = 0.69\,Tdn = 0.69 \times 471 = \mathbf{325}\ kN.$$

Allowable Tension, Ts = 285 kN

d) **To Determine Compressive Strength, C**

Member size: **ISA 100 100 10**; Length, l = 1000 mm (assumed)

Outstanding leg slenderness check:

$$b\,/\,t <= 10.5\sqrt{(250\,/\,Fy)} = 10.5\sqrt{(250\,/\,250)} = 10.5$$

b/t = 100/10 = 10 Compact section

Allowable Compressive strength, $C = 0.6\ fcd\ Ag = 0.6 \times 158 \times 1903/1000$

= 180 kN (IS 800:2007 – 11.3.1)

$$\mathbf{fcd} = \left(\mathbf{fy}\,/\,\gamma_{mo}\right)/\left[\phi + \sqrt{\left(\phi^2 - \lambda^2\right)}\right] = \left(250\,/\,1.1\right)/\left[0.92 + \sqrt{\left(0.92^2 - 0.757^2\right)}\right]$$

=**158** MPa (IS 800:2007 – 7.1.2.1)

λ = *slenderness ratio for single angle load through on leg*
(IS 800:2007 – 7.5.1.2)

$$\lambda = \sqrt{\left(\mathbf{k_1} + \mathbf{k_2}\lambda_{vv}{}^2 + \mathbf{k_3}\lambda_\phi{}^2\right)}^{\wedge 0.5}$$

$$= \sqrt{\left(0.2 + 0.35 \times 0.58^2 + 20 \times 0.113^2\right)} = \mathbf{0.757}$$

$$\lambda_{vv} = \left(\mathbf{l}\,/\,\mathbf{r_{vv}}\right)/\,\varepsilon.\sqrt{\left(\pi^2\mathbf{E}\,/\,\mathbf{250}\right)}$$

$$= \left(1000\,/\,19.4\right)/\left[1 \times \sqrt{\left(3.14^2 \times 200000\,/\,250\right)}\right] = 0.58$$

$$\lambda_\phi = \left(\mathbf{b1} + \mathbf{b2}\right)/\,\mathbf{2\,t}\,/\,\varepsilon.\left(\pi^2\mathbf{E}\,/\,\mathbf{250}\right)^{\wedge 0.5}$$

$$= \left(100 + 100\right)/\,2 \times 10\,/\left[1 \times \sqrt{\left(3.14^2 \times 200000\,/\,250\right)}\right]$$

$$= 0.113$$

$$\varepsilon = \left(\mathbf{250}\,/\,\mathbf{fy}\right)^{\wedge 0.5} = 1$$

$$\phi = \mathbf{0.5}\left[\mathbf{1} + \alpha\left(\lambda - \mathbf{0.2}\right) + \lambda^2\right]$$

$$= 0.5\left[1 + 0.49 \times \left(0.757 - 0.2\right) + 0.757^2\right] = 0.92$$

where,

	fixed	*hinged*
k1	0.2	0.7
k2	0.35	0.6
k3	20	5

$$r_{vv} = 19.4$$

$$\varepsilon = \sqrt{(250/250)} = 1$$

$$E = 200000\,\text{MPa}$$

$$b1 = b2 = 100\,\text{mm}$$

$$t = 10\,\text{mm}$$

$$\alpha = 0.49$$

Allowable comp., C = 180 kN

e) **To Determine Block Shear Capacity, Tb**

Member size: **ISA 100 100 10**

$t = 10$ mm; $b = 100$ mm; $(b_L = bs = b)$

Connection bolt: **4 nos M24**; $d = 24$ mm $d_0 = 27$ mm

Tdb = Smaller of Tdb1 or Tdb2 (IS 800:2007 – 6.4.1; 6.4.2)

$$\text{Tdb1} = \left[\text{Avg.fy} / \left(\sqrt{3}\,\gamma_{mo} \right) + 0.9\,A_{tn}\,\text{fu} / \gamma_{m1} \right]$$

Or

$$\text{Tdb2} = \left(0.9\,\text{Avn.fu} / \left(\sqrt{3}\,\gamma_{m1} \right) + \text{Atg fy} / \gamma_{mo} \right)$$

$$\text{Avg} = \text{gross area in shear along bolt line} = \left[44 + \left(3 \times 72 \right) \right] \times 10 = 2600\,\text{mm}^2$$

$$\text{Avn} = \text{net area in shear along bolt line} = 2600 - 4 \times 27 \times 10 = 1520\,\text{mm}^2$$

Atg = minimum gross area in tension perpendicular to the line of force

$$= \left(100 - 28.4 \right) \times 10 = 716\,\text{mm}^2$$

Atn = minimum net area in tension perpendicular to the line of force = $581\,\text{mm}^2$

$$\text{Tdb1} = \left[2600 \times 250 / \left(1.732 \times 1.1 \right) + 0.9 \times 581 \times 410 / 1.25 \right] / 1000 = 513\,\text{kN}$$

$$\text{Tdb2} = \left[0.9 \times 1520 \times 410 / (1.732 \times 1.25) + 716 \times 250 / 1.1 \right] / 1000 = 422 \text{ kN}$$

Tdb = 422 kN; Tb = 0.69 Tdb = 291 kN (IS 800:2007 – 7.5.1.2)

Block shear capacity, Tb = 291 kN > Ts; Okay.

Let us consider the maximum of tensile and compressive strength for a full strength joint design. Actual design value of F should be smaller of tension and compression force when length of member is known and compressive strength is calculated according to that length.

Full strength, F = 285 kN

f) Design of Connecting Bolts (working stress): (IS 800:2007 – 11.6.1)

Let us use **M24 HT Bolt 8.8 grade – Bearing type bolt**

Bolt diameter, d = 24 mm

γ_{mb} = 1.25 Material Safety factor.

Fub = 830 MPa Ultimate tensile stress of bolt.

Fu = 410 MPa Ultimate tensile stress of plate.

n_n = 0 No. of shear planes with threads intercepting the shear plane.

n_s = **1** No. of shear planes without threads intercepting the shear plane.

Asb = 453 mm^2 Nominal plain shank area of the bolt.

Anb = 353 mm^2 Net shear area of the bolt at thread.

e_1 = 44 mm End distance along bearing direction.

d_0 = 27 mm Diameter of the hole.

p = 60 mm Pitch distance along bearing direction (2.5d min).

kb = 0.491 [kb = smaller of e/3d0, (p/3d0 – 0.25), fub/fu, 1]; Refer Table 4.4.

Shear capacity:

Vnsb = (Fub/ $\sqrt{3}$) × (n_n. Anb + n_s. Asb) (IS 800:2007 – 10.3.3)

$$\text{Vnsb} = \left[(830 / 1.732) \times (1 \times 453) \right] / 1000 = 217 \text{ kN}$$

Permissible stress in bolt in shear, fasb = 0.6 Vnsb / Asb

$$= (0.6 \times 217 / 453) \times 1000 = 287 \text{ MPa.}$$

Bolt value, Vsb = Asb *x* fasb = 453 × 287/1000 = **130** kN (Single shear)

Bearing capacity:

Bearing Plate thickness = **10** mm [smaller of gusset and angle leg thickness].

Apb = 240 mm^2 bearing area on plate (d × t)

Vnpb = nominal bearing strength of a bolt = 2.5 kb.d.t.Fu

(IS 800:2007 – 10.3.4)

$$\text{Vpnb} = 2.5 \times 0.491 \times 24 \times 10 \times 410 / 1000 = 121 \, \text{kN}$$

Perm. Bearing stress in bolt/plate, fapb = 0.6 Vnpb/Apb

$$= (0.6 \times 121 / 240) \times 1000 = 303 \, \text{MPa}$$

Bolt value, Vsb = Apb. fapb = 240 × 303/1000 = **73** kN

Design force, **F = 285kN** (axial)
Shear Capacity of *a* bolt = 1 × 130 = 130 kN (single shear)

Bearing Capacity of a bolt = 73 kN

Bolt value or Design strength, Vsb = **73** kN
Number of bolts (N) = **4**
Total Capacity = 4 × 73 = **292** kN > **F = 285 kN; Safe**

4.4.5   Double Angle Sections – IS 800 (Working Stress)

a) Sketch

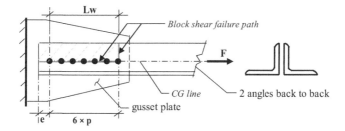

FIGURE 4.15   Double angle bolted joint – single row.

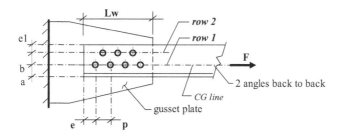

FIGURE 4.16   Double angle bolted joint – two rows.

b) Materials
Refer Table 4.6

c) **Design Parameters**

Refer Figure 4.13 for sectional view of the double angle member connected to the gusset plate.

Size of member: **2 - ISA 100 × 100 × 10**

Bolt: **M24** High tensile 8.8 grade

Nominal dia., d = 24 mm; No. of bolts = 7

Dia. of bolt hole, do = 27 mm; Gusset thickness, tg = 12 mm

Pitch distance, p = 72mm (3 × d)

Edge distance, e = 44 mm

b_L = bs = 100mm; t = 10 mm; Ag = 3806 mm²

An = 3536 mm²

w = 100 mm; w_1 = 28.4 mm

d) **To Determine Tensile Strength, Ts**:

Tensile strength, Ts shall be the lower value of tensile yielding in gross section (Ts1) and tensile rapture in net section (Ts2).

(i) Tensile yielding, Ts1 = Ag (0.6fy) = 3806 × (0.6 × 250)/1000 = **571** kN
(IS 800:2007 – 11.2.1)

(ii) Tensile rupture, Ts2 = 0.69 Tdn (IS 800:2007 – 11.2.1)

Tdn = 0.9 Anc fu/γ_{m1} + β Ago fy/γ_{mo} (IS 800:2007 – 6.3.3)

$$\beta = 1.4 - 0.076 \left(w/t \right) \left(fy/fu \right) \left(bs/Lc \right) <= 0.9\, fu\, \gamma_{mo}/fy\, \gamma_{ml} >= 0.7$$

where,

w = outstanding leg width = 100 mm (IS 800:2007 – 6.3.4)

bs = shear lag width = w + w1 = 100 + 28.4 = 128.4 mm

Lc = length of end connection = distance between two outermost bolts
= (7 – 1) × 72 = 432 mm

$$Anc = \text{net area of the connected leg} = 1460\, mm^2$$

$$Ago = \text{gross area of outstanding leg} = 2000\, mm^2$$

t = thickness of the leg = 10 mm

β = lower of the following values of (i) and (ii) but more than or equal to 0.7

(iii) $1.4 - 0.076 \left(w/t \right) \left(fy/fu \right) \left(bs/Lc \right)$
$= 1.4 - 0.076 \left(100/10 \right) \times \left(250/410 \right) \times \left(128.4/432 \right) = 1.262$

(iv) $0.9\, fu\, \gamma_{mo}/fy\, \gamma_{ml} = 0.9 \times 410 \times 1.1 / \left(250 \times 1.25 \right) = 1.299$

(v) ≥ 0.7

So,

$$\beta = \mathbf{1.262}$$

Now,

$$\text{Tdn} = 0.9 \times 1460 \times 410 / 1.25 + 1.262 \times 2000 \times 250 / 1.1 = 1004628 \text{ N}$$

$$= 1005 \text{ kN}$$

$$\text{Ts2} = 0.69 \text{ Tdn} = 0.69 \times 1005 = \mathbf{693} \text{ kN.}$$

Allowable Tension, Ts = 571 kN

e) **To Determine Compressive Strength, C**
   Member size: **ISA 100 100 10**

TABLE 4.11
Sectional Properties of Double Angle Member

Member Section			For Double Legged Angle Member							
Section Provided:		S mm	wt/m (kg)	A_g (mm^2)	r_x (mm)	r_y (mm)	r_{vv} (mm)	I_x (mm^4)	I_y (mm^4)	C_{yy} (mm)
2	ISA 100 100 10	12	29.8	3806	30.50	46.0	19.4	3540000	8043868	28.4

Single angle sizes							
b	d	t		Connection bolt			An
mm	mm	mm	nos	φ dia.	φ hole	rows	mm^2
100	100	10	7	24	27	1	3536

where, An = net area after deduction of bolt holes.

$$\text{An} = \text{Ag} - \left(\text{no of rows} \times \phi \text{ hole} \times t\right)$$

Buckling class
   Buckling class = c for buckling about any axis (IS 800:2007 – Table 10)
   Imperfaction factor, $\alpha = 0.49$ (IS 800:2007 – Table 7)
Proportioning limit

$$\varepsilon = \sqrt{(250 / \text{fy})} = \sqrt{(250 / 250)} = 1.00$$

Outstanding leg

b/t = 100/10 = 10 Compact (IS 800:2007 – 3.7.2)
Limiting width to thickness ratio:
Plastic, $\lambda r = 9.4\varepsilon = 9.4 \times 1 = 9$
Compact, $\lambda p = 10.5\varepsilon = 10.5 \times 1 = 10.5$
Semi compact, $\lambda sp = 15.7\ \varepsilon = 15.7 \times 1 = 15.7$

Effective Lengths (L) and Slenderness ratio (L/r):

Laterally Unsupported length: Lx = 1 M(*) Kx = 0.85; KLx = 0.85 M
Ly = 1 M; Ky = 1; KLy = 1.00 M

Maximum slenderness ratio (L/r) shall not exceed **250** (IS 800:2007 – Table 3)

Here, KLx / rx = 850 / 0 30.5 = 28
KLy/ry = 1000/46 = 22   **L/r = 28** < 250; Okay.
[(*) Length assumed to get full strength of member]

Sectional capacity / Design strength:
Axial compression

(i) *Compression due to buckling about major axis, z-z [Lz / rz (=rx)]*
(IS 800:2007 – 7.5.2.1)

Design compression strength, Pdz = Ae. Fcd (IS 800:2007 – 7.1.2)
Where, Ae = effective sectional area
And fcd = design compressive stress

$$fcd = \chi\, fy\, /\, \gamma_{mo} \quad <= fy\, /\, \gamma_{mo}$$

$$\chi = 1 / \left[\phi + \sqrt{\left(\phi^2 - \lambda^2 \right)} \right]$$

$$\phi = 0.5 \left[1 + \alpha\left(\lambda - 0.2 \right) + \lambda^2 \right]$$

$$\lambda = \sqrt{\left(fy\, /\, fcc \right)}$$

fcc = $\pi^2 E/(KLz/rz)^2$ Euler buckling stress
Now, putting following values into above equation
$\alpha = 0.49$; E = 200000 MPa; fy = 250 MPa
γ_{mo} = 1.10 KLx/r_x = 28 Ae = **3806** mm^2 = Ag

$$fcc = 3.14^2 \times 200000 / 28^2 = 2515\, MPa$$

$$\lambda = \sqrt{\left(250 / 2515 \right)} = 0.32$$

$$\phi = 0.5\left[1 + 0.49\left(0.32 - 0.2\right) + 0.32^2\right] = 0.58$$

$$\chi = 1/\left[0.58 + \sqrt{\left(0.58^2 - 0.32^2\right)}\right] = 0.94$$

fcd = 0.94 × 250/1.1 = 214 MPa < = fy/γ_{mo} < = 250/1.1

<= 227 MPa

fcd = **214** MPa $\left(= fcdx\right)$

Pdx = Ae.fcd = 3806 × 214 / 1.1 = **814 kN**

(ii) *Compression due to buckling about major axis, z-z [Lz / rz (=rx)]*
   (IS 800:2007 – 7.5.2.1)
      Design compression strength, Pdz = Ae. fcd (IS 800:2007 – 7.1.2)
      Where, Ae = effective sectional area
      And fcd = design compressive stress

$$fcd = \chi \; fy \; / \; \gamma_{mo} \quad <= fy \; / \; \gamma_{mo}$$

$$\chi = 1/\left[\phi + \sqrt{\left(\phi^2 - \lambda^2\right)}\right]$$

$$\phi = 0.5\left[1 + \alpha\left(\lambda - 0.2\right) + \lambda^2\right]$$

$$\lambda = \sqrt{\left(fy \; / \; fcc\right)}$$

fcc = $\pi^2 E/(KLz/rz)^2$ Euler buckling stress
Now, putting following values into above equation
   α = 0.49 E = 200000 MPa fy = 250 MPa
   γ_{mo} = 1.10 KLy/r_y = 22 Ae = **3806** mm² = Ag

$$fcc = 3.14^2 \times 200000 / 22^2 = 4074 \; MPa$$

$$\lambda = \sqrt{\left(250 / 4074\right)} = 0.25$$

$$\phi = 0.5\left[1 + 0.49\left(0.25 - 0.2\right) + 0.25^2\right] = 0.54$$

$$\chi = 1/\left[0.54 + \sqrt{\left(0.54^2 - 0.25^2\right)}\right] = 0.982$$

fcd = 0.982 × 250/1.1 = 223 MPa; < = fy/γ_{mo}; < = 250/1.1

<= 227 MPa

$$\text{fcd} = \mathbf{214}\,\text{MPa}\left(= \text{fcdy}\right)$$

$$\mathbf{Pdy} = \text{Ae.fcd} = 3806 \times 223 / 1.1 = \mathbf{849\,kN}$$

The design compressive strength, Pd will be lowest of above calculated strengths of the member.

$$\mathbf{Pd = 814\,kN}$$

Permissible compressive strength = 0.6 Pd = 0.6 × 814 = **488 kN** (working stress)

Allowable compression, C = 488 kN
f) **To Determine Block Shear Capacity, Tb:**
Member size**: 2 - ISA 100 100 10**
t = 10 mm; b = 100 mm; $(b_L = \text{bs} = b)$
Connection bolt: **7 nos M24**; d = 24 mm; d_0 = 27 mm
Tdb = Smaller of Tdb1 or Tdb2 (IS 800:2007 – 6.4.1; 6.4.2)

$$\text{Tdb1} = \left[\text{Avg.fy} / \left(\sqrt{3}\,\gamma_{mo}\right) + 0.9\,A_{tn}\text{fu} / \gamma_{m1}\right]$$

Or

$$\text{Tdb2} = \left(0.9\text{Avn.fu} / \left(\sqrt{3}\gamma_{m1}\right) + \text{Atg fy} / \gamma_{mo}\right)$$

Avg = gross area in shear along bolt line = $2\left[44 + \left(6 \times 72\right)\right] \times 10 = 9520\,\text{mm}^2$

Avn = net area in shear along bolt line = $9520 - 7 \times 27 \times 10 = 7630\,\text{mm}^2$

Atg = minimum gross area in tension perpendicular to the line of force

$$= 2\left(100 - 28.4\right) \times 10 = 1432\text{mm}^2$$

Atn = minimum net area in tension perpendicular to the line of force = $1162\,\text{mm}^2$

$$\text{Tdb1} = \left[9520 \times 250 / \left(1.732 \times 1.1\right) + 0.9 \times 1162 \times 410 / 1.25\right] / 1000 = 1592\,\text{kN}$$

$$\text{Tdb2} = \left[0.9 \times 7630 \times 410 / \left(1.732 \times 1.25\right) + 1432 \times 250 / 1.1\right] / 1000 = 1626\,\text{kN}$$

Tdb = 1592 kN; **Tb** = 0.69 Tdb = **1099** kN (IS 800:2007 – 7.5.1.2)

Block shear capacity, Tb = 1099 kN > Ts; Okay.
Let us consider the maximum of tensile and compressive strength for a full strength joint design. Actual design value of F should be smaller of ten-

sion and compression force when length of member is known and compressive strength is calculated according to that length.

Full strength, F = 571 kN

g) **Design of Connecting Bolts (working stress)**: (IS 800:2007 – 11.6.1)
Let us use,

M24	HT Bolt 8.8 grade – Bearing type bolt
Bolt diameter	d = 24 mm
$\gamma_{mb} = 1.25$	Material Safety factor.
Fub = 830 MPa	Ultimate tensile stress of bolt.
Fu = 410 MPa	Ultimate tensile stress of plate.
$n_n = 0$	No. of shear planes with threads intercepting the shear plane.
$n_s = 2$	No. of shear planes without threads intercepting the shear plane.
Asb = 453 mm²	Nominal plain shank area of the bolt.
Anb = 353 mm²	Net shear area of the bolt at thread.
$e_1 = 44$ mm	End distance along bearing direction.
$d_0 = 27$ mm	Diameter of the hole.
p = 72 mm	Pitch distance along bearing direction (3 d min).
kb = 0.543	[kb = smaller of e/3d0, (p/3d0 – 0.25), fub/fu, 1]; Refer Table 4.4.

Shear capacity:

Vnsb = (Fub/ $\sqrt{3}$) × (n_n. Anb + n_s. Asb) (IS 800:2007 – 10.3.3)

$$Vnsb = \left[(830/1.732)\times(2\times453)\right]/1000 = 434 \text{ kN}$$

Bolt value, Vsb = 0.6 Vnsb = 0.6 × 434 = **260** kN (Double shear)

Bearing capacity:

Bearing Plate thickness = **12** mm; [smaller of the sum of double leg thickness and gusset plate thickness].

Apb = 24 × 12 = 288 mm² bearing area on plate (d × t)
Vnpb = nominal bearing strength of a bolt = 2.5 kb .d .t. Fu
(IS 800:2007 – 10.3.4)

$$Vpnb = 2.5\times0.543\times24\times12\times410/1000 = 160 \text{ kN}$$

Bolt value, Vsb = 0.6 Vpnb = 0.6 × 160 = **96** kN

Design force, **F = 571 kN** (axial)
Shear Capacity of *a* bolt = 2 × 130 = 260 kN (double shear)

$$\text{Bearing Capacity of a bolt} = 96 \text{ kN}$$

Bolt value or Design strength, Vsb = **96** kN per bolt
Number of bolts (N) = **7**
Total Capacity = 7 × 96 = **672** kN   > F = **571 kN; Safe**

4.5   BRACING MEMBER JOINTS

In this chapter, the workout examples are provided for bracing members joined with HT bolts bearing the applied load.

4.5.1   TUBULAR SECTION

a) Sketch

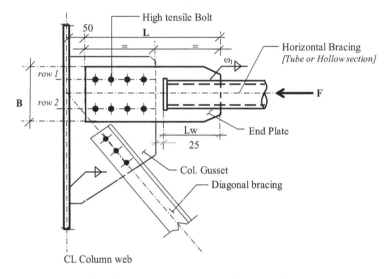

Elevation

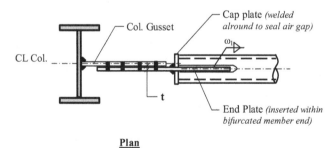

Plan

FIGURE 4.18   Bracing end connection.

b) Material

TABLE 4.12
Properties of Hollow Sections

IS 10748

Type		2	3		
Yield stress	fy	210	240		MPa
d or t					
Tensile stress	fu	330	410		MPa
Plates E 250 (Fe 410 W)		A	B	C	
Yield stress	fy	250	240	230	MPa
d or t		<20	20–40	>40	
Tensile stress	fu	410	410	410	MPa
Elasticity	E			200000	MPa
	γ_{m0}	1.10	γ_{m1}	1.25	
Material safety factor	Bolt friction type		Bolt bearing type		
	γ_{mf}	1.25	γ_{mb}	1.25	for bolt
HT Bolt					
8.8 (d ≤ 16)	Fyb	640	Fub	800	Mpa
8.8 (d > 16)	Fyb	660	Fub	830	Mpa
Electrode Ex40-44XX:		fuw	540	Mpa	
γ_{mw}	1.25	fyw	330	Mpa	

c) **Design Parameters**
Member: Tube 125L
Outside dia, OD = 139.7 mm; Pipe thickness, Th = 4.5 mm
Area of cross section, A = 1910 mm^2; Nominal weight = 15 Kg/m
End plate: L = 560 mm; B = 280 mm; t = 12 mm
Column Gusset plate thickness, tg = 12 mm
HT bolts in row 1 (= row2) = 4; Total no. of bolts = 8
Bolt size: M16; bolt dia.= 16 mm; hole dia.= 18 mm
Size of fillet weld, ω_1 = 6 mm (shop)

d) **End Plate sizing**

$$\text{Area of cross section of member} = 1910 \, \text{mm}^2$$

Required cross sectional area of end plate along force, **Arqd** = 1.05 × 1910
= 2006 mm^2
(Assumed 5% in excess of area of member joined.)
Effective area of cross section of end plate = 280 × 12 – 2 × 18 × 12 = 2928
mm^2 > Arqd; Safe.

e) To calculate allowable force in the member, F

TABLE 4.13
Allowable Comp Strength [50L]

IS 800 2007 – 7.1.2.1

Section 125L

r min	47.8	mm	
K	0.85		
L	1000	mm	
α	0.34		Imperfection factor a (IS: 800 Table 7 and Table 10)
γ_{mo}	1.10		
fy	210	Mpa	
E	200000	Mpa	
fy/γ_{mo}	191		
$\lambda =$	0.1835		
$\phi =$	0.514		
fcd =	191	Mpa	fcd = design compressive stress [limit state]
IS 800 2007 – 11.3.1			
fabc =	**115**	Mpa	fabc = 0.6 fcd, design compressive stress [working stress]

Allowable Compressive strength, **C** = fabc. *A* = 115 × 1910/1000 = **220 kN**

Allowable Tensile strength, Ts = 0.6 fy. Ag = 0.6 × 210 × 1910/1000 = **241 kN**

Allowable Block shear capacity, Tb:

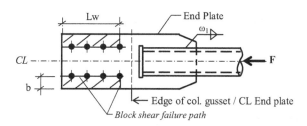

FIGURE 4.19 Block shear failure.

Member: **Tube 125 L**
End plate:
L = 560 mm
B = 280 mm
t = 12 mm

Connection bolt: M16
d = 16 mm
do = 18 mm
Nos. per row = 4
Pitch distance = 3d = 48 mm
Edge dist, e_0 = 29 mm

$$Lw = (4-1) \times 48 + 29 = 173 \, mm$$

$$b = e_0 = 29 \, mm$$

Tdb = Smaller of Tdb1 or Tdb2 (IS 800:2007 – 6.4.1; 6.4.2)

$$Tdb1 = \left[Avg.fy / (\sqrt{3}\gamma_{mo}) + 0.9 \, A_{tn}fu / \gamma_{m1} \right]$$

Or

$$Tdb2 = \left(0.9Avn.fu / (\sqrt{3}\gamma_{m1}) + Atg \, fy / \gamma_{mo} \right)$$

Avg = gross area in shear along bolt line = $2 \times 173 \times 12 = 4152 \, mm^2$

Avn = net area in shear along bolt line = $4152 - 2 \times 4 \times 18 \times 12 = 2424 \, mm^2$

Atg = minimum gross area in tension perpendicular to the line of force

$$= 2 \times 29 \times 12 = 696 \, mm^2$$

Atn = minimum net area in tension perpendicular to the line of force = $480 \, mm^2$

$$Tdb1 = \left[4152 \times 250 / (1.732 \times 1.1) + 0.9 \times 480 \times 410 / 1.25 \right] / 1000 = 687 \, kN$$

$$Tdb2 = \left[0.9 \times 2424 \times 410 / (1.732 \times 1.25) + 696 \times 250 / 1.1 \right] / 1000 = 414 \, kN$$

Tdb = 414 kN; Tb = 0.69 Tdb = 285kN (IS 800:2007 – 7.5.1.2)

Block shear capacity, Tb = 285 kN > Ts; Ok.
 Let us consider the maximum of tensile and compressive strength for a full strength joint design. Actual design value of F should be smaller of tension and compression force, when length of member is known and compressive strength is calculated according to that length.
 Design force, F = 241 kN
f) **Strength of fillet weld, ω_1**
 See Figure 4.18.
 L = 560 mm; B = 280 mm
 Space between edge of col gusset and member end = 25 mm

$$Weld \, length = 0.5 \times 560 - 25 = 255 \, mm \, / \, side$$

ω_1 = 6 mm; fyw = 330 MPa (yield stress); faw = 132 MPa
as per IS 800:2007 Cl 10.5.7.1,

$$fwn = fuw / \sqrt{3} = 540 / 1.732 = 312 \text{ MPa}$$

as per IS 800:2007 Cl 11.6.2.4
The permissible stress of weld, faw = 0.6 fwn = 0.6 × 312 = 187 MPa
as per IS 800:2007 Cl 11.6.3.1
The permissible stress of weld, faw = 0.4 fyw = 0.4 × 330 = **132** MPa
Let us consider, **fwd = 132 MPa**
Strength of fillet weld, ω_1 (Shop weld)

= Shear strength in weld (along the applied force, F)

+ Tensile resistance in weld (perpendicular to the axis of force)

$$= 0.7.\omega_1.fwd \times \mathbf{4}\,Lw + 0.7\,\omega_1.(\text{pipe OD}+12).(0.5\,fwd) \quad \text{Mpa}$$

$$= 0.7 \times 6 \times 132 \times (4 \times 255) / 1000 + 0.7 \times 6 \times 152 \times (0.5 \times 132) / 1000$$

= **610** kN   **> F = 241 kN; Safe**

g) Connection bolt (IS 800:2007 – 11.6.1)
Let us use,

M16	HT Bolt 8.8 grade – Bearing type bolt
Bolt diameter	d = 16 mm
γ_{mb} = 1.25	Material Safety factor
Fub = 800 MPa	Ultimate tensile stress of bolt.
Fu = 410 MPa	Ultimate tensile stress of plate.
n_n = 0	No. of shear planes with threads intercepting the shear plane.
n_s = 1	No. of shear planes without threads intercepting the shear plane.
Asb = 201 mm²	Nominal plain shank area of the bolt.
Anb = 157 mm²	Net shear area of the bolt at thread.
e_1 = 29 mm	End distance along bearing direction.
d_0 = 18 mm	Diameter of the hole.
p = 74 mm	pitch distance along bearing direction (3 d min)
	[$p = (0.5 \times 560 - 2 \times 29)/(4 - 1)$]
kb = 0.537	[kb = smaller of e/3d0, (p/3d0 – 0.25), fub/fu, 1]; Refer Table 4.4

Shear capacity:
Vnsb = (Fub/ $\sqrt{3}$) × (n_n. Anb + n_s. Asb) (IS 800:2007 – 10.3.3)

$$Vnsb = \left[(800 / 1.732) \times (1 \times 201)\right] / 1000 = 93 \text{ kN}$$

Bolt value, Vsb = 0.6 Vnsb = 0.6 × 93 = **56** kN (single shear)

Bearing capacity:

Bearing Plate thickness = **12** mm [Smaller of end plate and gusset plate thickness].

Apb = 16 × 12 = 192 mm² bearing area on plate (d × t)

Vnpb = nominal bearing strength of a bolt = 2.5 kb .d .t. Fu

(IS 800:2007 – 10.3.4)

$$\text{Vpnb} = 2.5 \times 0.537 \times 16 \times 12 \times 410 / 1000 = 106 \text{ kN}$$

Bolt value, Vsb = 0.6 Vpnb = 0.6 × 106 = **64** kN

Design force, **F = 241 kN** (axial)

Shear Capacity of a bolt = 56 kN

Bearing Capacity of a bolt = 64 kN

Bolt value or Design strength, Vsb = **56** kN

Number of bolts (row 1 + 2) = 8

Total Capacity = 8 × 56 = **448** kN > **F = 241 kN; Safe**.

4.5.2   Hollow Rectangular Sections

a) Sketch

Please see Figure 4.18

b) Materials

Refer Table 4.11

c) Design Parameters

Member: Hollow section 172 92 4.8; Nominal weight = 15 Kg/m

Outside depth, D = 172 mm

Wide, B = 92 mm

Thickness, T = 4.8 mm

Area of cross section, A = 2383 mm²

End plate: L = 690 mm; B = 310 mm; t = 12 mm

Column Gusset plate thickness, tg = 12 mm

HT bolts in row 1 (= row 2) = 4; Total no. of bolts = 8

Bolt size: **M16**; bolt dia. = 16 mm; Hole dia. = 18 mm

Size of fillet weld, ω_1 = **5** mm (shop)

d) End Plate Sizing

$$\text{Area of cross section of member} = 2383 \text{ mm}^2$$

Required cross sectional area of end plate along force, **Arqd** = 1.05 × 2383 = 2502 mm²

(Assumed 5% in excess of area of member joined.)

Effective area of cross section of end plate = $310 \times 12 - 2 \times 18 \times 12 = 3288 \text{ mm}^2$

> **Arqd; Safe**.

e) **To Calculate Allowable Force in the Member, F**

TABLE 4.14
Allowable Comp Strength

IS 800:2007 – 7.1.2.1

Section 172 × 92 × 4.8

r min	62	mm	
K	0.85		
L	1000	mm	
α	0.34		Imperfection factor a (IS: 800 Table 7 and Table 10)
γ_{mo}	1.10		
Fy	210	MPa	
E	200000	MPa	
fy/γ_{mo}	191		
λ =	0.1415		
φ =	0.5001		
fcd =	191	MPa	fcd = design compressive stress [limit state]
IS 800:2007 – 11.3.1			
fabc =	**115**	MPa	fabc = 0.6 fcd, design compressive stress [working stress]

Allowable **Compressive strength, C** = fabc. A = 115 × 2383/1000 = **274 kN**

Allowable **Tensile strength, Ts** = 0.6 fy. Ag = 0.6 × 210 × 2383/1000 = **300 kN**

Allowable Block shear capacity, Tb:
Refer Figure 4.19
Member: Rectangular hollow section 172 92 4.8
End plate:
L = 690 mm; B = 310 mm; t = 12 mm
Connection bolt: M16
d = 16 mm; do = 18 mm; Nos. per row = 4
Pitch distance = 3d = 48 mm; Edge dist., e_0 = 29 mm

$$Lw = (4-1) \times 48 + 29 = 173\,mm$$

$$b = e_0 = 29\,mm$$

Tdb = Smaller of Tdb1 or Tdb2 (IS 800:2007 – 6.4.1; 6.4.2)

$$Tdb1 = \left[Avg.fy / \left(\sqrt{3}\,\gamma_{mo} \right) + 0.9\,A_{tn}fu / \gamma_{m1} \right]$$

Or

$$Tdb2 = \left(0.9 Avn.fu / \left(\sqrt{3}\gamma_{m1}\right) + Atg\, fy / \gamma_{mo}\right)$$

Avg = gross area in shear along bolt line = $2 \times 173 \times 12 = 4152\,mm^2$

Avn = net area in shear along bolt line = $4152 - 2 \times 4 \times 18 \times 12 = 2424\,mm^2$

Atg = minimum gross area in tension perpendicular to the line of force
$= 2 \times 29 \times 12 = 696\,mm^2$

Atn = minimum net area in tension perpendicular to the line of force = $480\,mm^2$

$$Tdb1 = \left[4152 \times 250 / (1.732 \times 1.1) + 0.9 \times 480 \times 410 / 1.25\right] / 1000 = 687\,kN$$

$$Tdb2 = \left[0.9 \times 2424 \times 410 / (1.732 \times 1.25) + 696 \times 250 / 1.1\right] / 1000 = 571\,kN$$

Tdb = 571 kN; Tb = 0.69 Tdb = 394kN (IS 800:2007 – 7.5.1.2)
Block shear capacity, Tb = 394 kN > Ts; Okay.
Let us consider the maximum of tensile and compressive strength for a full strength joint design. Actual design value of F should be smaller of tension and compression force when length of member is known and compressive strength is calculated according to that length.
Design force, F = 300 kN
f) **Strength of Fillet Weld, ω_1**
See Figure 4.18 above.
L = 690 mm; B = 310 mm
Space between edge of col gusset and member end = 25 mm
Weld length = $0.5 \times 690 - 25 = 320\,mm$ / side
$\omega_1 =$ 5 mm; fyw = 330 MPa (yield stress); faw = 132 MPa
as per IS 800:2007 Cl 10.5.7.1,
fwn = fuw / $\sqrt{3}$ = 540 / 1.732 = 312 MPa
as per IS 800:2007 Cl 11.6.2.4
The permissible stress of weld, faw = 0.6 fwn = 0.6 × 312 = 187 MPa
as per IS 800:2007 Cl 11.6.3.1
The permissible stress of weld, faw = 0.4 fyw = 0.4 × 330 = **132** MPa
Let us consider, **fwd = 132 MPa**
Strength of fillet weld, ω_1 (Shop weld)

= Shear strength in weld $\left(\text{along the applied force}, F\right)$ + Tensile resistance in weld $\left(\text{perpendicular to the axis of force}\right)$

$= 0.7.\omega_1.fwd \times 4\,Lw + 0.7\omega_1.\left(\text{pipe OD} + 12\right).\left(0.5\,fwd\right) MPa$

$= 0.7 \times 5 \times 132 \times \left(4 \times 320\right) / 1000 + 0.7 \times 5 \times 104 \times \left(0.5 \times 132\right) / 1000$

$= 615\,kN \quad > F = 300\,kN; Safe$

g) Connection Bolt (IS 800:2007 – 11.6.1)

Let us use,

M16	HT Bolt 8.8 grade – Bearing type bolt
Bolt diameter	d = 16 mm
$\gamma_{mb} = 1.25$	Material Safety factor.
Fub = 800 MPa	Ultimate tensile stress of bolt.
Fu = 410 MPa	Ultimate tensile stress of plate.
$n_n = 0$	No. of shear planes with threads intercepting the shear plane.
$n_s = 1$	No. of shear planes without threads intercepting the shear plane.
Asb = 201 mm^2	Nominal plain shank area of the bolt.
Anb = 157 mm^2	Net shear area of the bolt at thread.
$e_1 = 29$ mm	End distance along bearing direction.
$d_0 = 18$ mm	Diameter of the hole.
p = 95.67 mm	Pitch distance along bearing direction (3 d min)
	$[p = (0.5 \times 690 - 2 \times 29)/(4 - 1)]$.
kb = 0.537	[kb = smaller of e/3d0, (p/3d0 – 0.25), fub/fu,1]; Refer Table 4.4.

Shear capacity:

Vnsb = (Fub/ $\sqrt{3}$) × (n_n. Anb + n_s. Asb) (IS: 800 – 2007 – 10.3.3)

$$\text{Vnsb} = \left[\left(800 / 1.732 \right) \times \left(1 \times 201 \right) \right] / 1000 = 93 \text{ kN}$$

Bolt value, Vsb = 0.6 Vnsb = 0.6 × 93 = **56** kN (single shear)

Bearing capacity:

Bearing Plate thickness = **12** mm [Smaller of end plate and gusset plate thickness].

Apb = 16 × 12 = 192 mm^2 bearing area on plate (d × t)

Vnpb = nominal bearing strength of a bolt = 2.5 kb .d .t. Fu
(IS:800 – 2007 – 10.3.4)

$$\text{Vpnb} = 2.5 \times 0.537 \times 16 \times 12 \times 410 / 1000 = 106 \text{ kN}$$

Bolt value, Vsb = 0.6 Vpnb = 0.6 × 106 = **64** kN
Design force, **F = 300 kN** (axial)
Shear Capacity of a bolt = 56 kN
Bearing Capacity of a bolt = 64 kN
Bolt value or Design strength, Vsb = **56** kN
Number of bolts (row 1 + 2) = 8
Total Capacity $= 8 \times 56 = $ **448** kN $> F = $ **300 kN; Safe.**

4.6   SPLICE JOINT – PLATED AND ROLLED SECTION

Following sketches show the elevations, sectional views of a beam splice joint. Separate views are given showing different group of bolts in web splice joint, which are used in the design examples and charts provided in this book. Designer may create alternate pattern based on load and member size.

a) Elevation and Sections

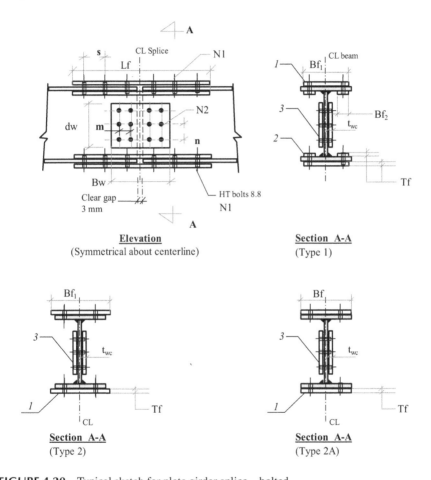

FIGURE 4.20   Typical sketch for plate girder splice – bolted.

b) Groups of Bolt for Web Joint

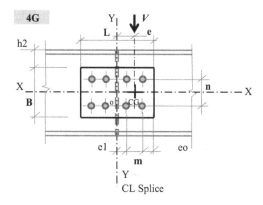

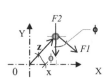

$F_1 = Ve \cdot z / \sum (x^2 + y^2) \quad I = \sum (x^2 + y^2)$
$F2 = V / n_o$
$F = \sqrt{(F_1^2 + F_2^2 + 2\, F_1\, F_2 \cos\phi)}$

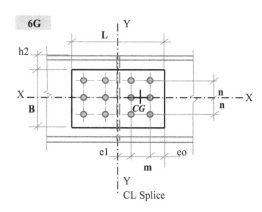

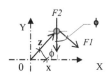

$F_1 = Ve \cdot z / \sum (x^2 + y^2) \quad I = \sum (x^2 + y^2)$
$F2 = V / n_o$
$F = \sqrt{(F_1^2 + F_2^2 + 2\, F_1\, F_2 \cos\phi)}$

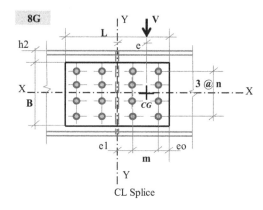

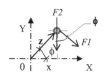

$F_1 = Ve \cdot z / \sum (x^2 + y^2) \quad I = \sum (x^2 + y^2)$
$F2 = V / n_o$
$F = \sqrt{(F_1^2 + F_2^2 + 2\, F_1\, F_2 \cos\phi)}$

10G

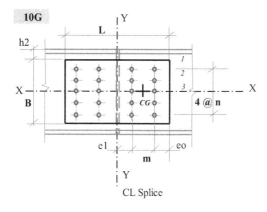

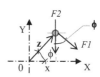

$F_1 = Ve . z / \sum (x^2 + y^2)$ $I = \sum (x^2 + y^2)$
$F2 = V / n_o$
$\mathbf{F = \sqrt{(F_1^2 + F_2^2 + 2\ F_1\ F_2\ \cos\phi)}}$

15g

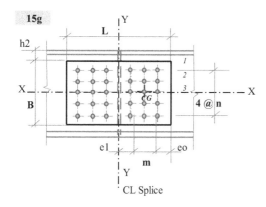

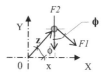

$F_1 = Ve . z / \sum (x^2 + y^2)$ $I = \sum (x^2 + y^2)$
$F2 = V / n_o$
$\mathbf{F = \sqrt{(F_1^2 + F_2^2 + 2\ F_1\ F_2\ \cos\phi)}}$

12G

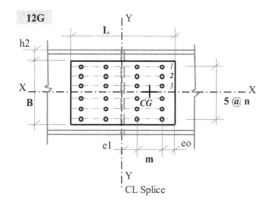

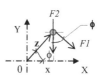

$F_1 = Ve . z / \sum (x^2 + y^2)$ $I = \sum (x^2 + y^2)$
$F2 = V / n_o$
$\mathbf{F = \sqrt{(F_1^2 + F_2^2 + 2\ F_1\ F_2\ \cos\phi)}}$

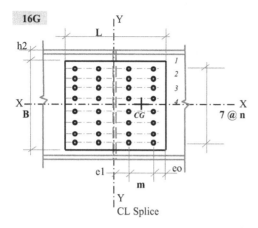

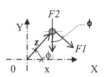

$$F_1 = Ve \cdot z \,/\, \sum (x^2 + y^2) \quad I = \sum (x^2 + y^2)$$
$$F2 = V \,/\, n_o$$
$$F = \surd\,(F_1{}^2 + F_2{}^2 + 2\,F_1\,F_2\,\cos\phi)$$

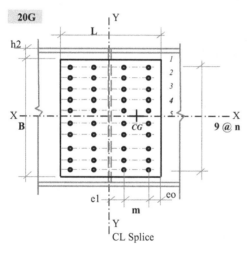

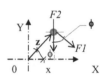

$$F_1 = Ve \cdot z \,/\, \sum (x^2 + y^2) \quad I = \sum (x^2 + y^2)$$
$$F2 = V \,/\, n_o$$
$$F = \surd\,(F_1{}^2 + F_2{}^2 + 2\,F_1\,F_2\,\cos\phi)$$

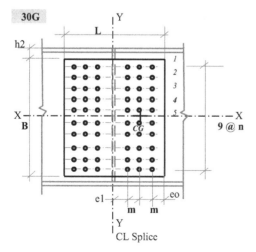

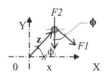

$$F_1 = Ve \cdot z \,/\, \sum (x^2 + y^2) \quad I = \sum (x^2 + y^2)$$
$$F2 = V \,/\, n_o$$
$$F = \surd\,(F_1{}^2 + F_2{}^2 + 2\,F_1\,F_2\,\cos\phi)$$

FIGURE 4.21–4.29   Group of bolts for web joint.

4.6.1 BEAMS AND GIRDER WITH BEARING TYPE BOLT – IS 800 (WORKING STRESS)

a) **Sketch**

Please see Figure 4.20.

b) **Materials**

Same as in Table 4.6.

c) **Member Descriptions**

Spliced member: **ISMB 600 Type – 2**

Flange:	*Web*:
wide, bf = 210 mm	depth, d = 600 mm
thickness, tf = 20.8 mm	thickness, tw = 12 mm

Connection Bolt: **M20**; HT Bolt 8.8 grade – bearing type

d) **Design Parameters**

1. Flange force, F: Full tension capacity = 0.6 fy. Af; where, Af = flange area
2. Web shear, V: 50% of web shear capacity (= 0.4 fy Aw); where, Aw = web area
3. Sum area of cover plates is 5% more than the area of section spliced.
4. All bolts are 8.8 grade HT bolts bearing type.

e) **Sizing of Flange Cover Plate and Connection Bolt**

Let us consider following dimensions for flange cover plates:

Flange cover plate -1:

wide, Bf1 = 230 mm; thick, Tf = 20 mm; length, Lf = 600 mm

No. of plate per flange = 1

Flange cover plate 2 is not used for Type 2 joint.

Cross sectional area of girder flange, Af = 210 × 20.8 = 4368 mm^2

Sectional area of cover plate, A required = Af × 1.05 = 4368 × 1.05 = 4586 mm^2

Total sectional area of cover plates, A provided = 1 × 230 × 20 = 4600 mm^2 > A required; Okay.

Flange connection bolts, N1:

Bolt: M20; HT 8.8 grade, bearing type

Bolt dia., d = 20 mm

Hole dia., do = 22 mm

Edge dist., e_0 = 32 mm

No. of rows = 2

Nos of Bolt/row = 4

$$Lw = (600/2) - 32 = 268 \, mm$$

$$\text{Spacing of bolt,} s = (0.5 \times 600 - 32 \times 2)/(4-1) = 78.67 \, mm > 2.5 \, d; Okay.$$

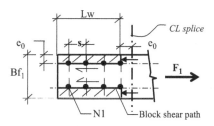

FIGURE 4.30   Plan view of top cover plate 1.

Tension in cover plate:
The actual tensile stress on the cover plate shall be smaller of the following:
a) Yielding of gross section, fat = 0.6 fy (IS 800:2007 – 11.2.1)
b) Rupture of the net section, fat = 0.69 Tdn/Ag
c) Block shear resistance, fat = 0.69 Tdb/Ag

a) by yielding:

$$\text{fat} = 0.6 \times 250 = \mathbf{150}\,\text{MPa}$$

b) by rapture: (IS 800:2007 – 6.3.1)

$$\text{An} = 230 \times 20 - 2 \times 22 = 4556\,\text{mm}^2$$

$$\text{Tdn} = 0.9\,\text{An fu} / \gamma_{m1} = 0.9 \times 4556 \times 410 / 1.25 / 1000 = 1345\,\text{kN}$$

$$\text{fat} = 0.69 \times 1345000 / 4556 = \mathbf{204}\,\text{MPa}$$

c) by block shear:
Cover plate 1: Lw = 268mm; $b = e_0 = 32$ mm
   t = 20 mm
   Tb = Smaller of Tdb1 or Tdb2 (IS 800:2007 – 6.4.1; 6.4.2)

$$\boldsymbol{Tdb1 = \left[Avg.fy \,/ \left(\sqrt{3}\, \gamma_{mo} \right) + 0.9\, Atn\, fu \,/\, \gamma_{m1} \right]}$$

$$\boldsymbol{Tdb2 = \left(0.9 A_{vn} fu \,/ \left(\sqrt{3}\, \gamma_{m1} \right) + A_{tg}\, fy \,/\, \gamma_{mo} \right)}$$

Avg = gross area in shear along bolt line $= 2 \times 268 \times 20 = 10720\,\text{mm}^2$

Avn = net area in shear along bolt line $= 10720 - 2 \times 4 \times 22 \times 20 = 7200\,\text{mm}^2$

Atg = minimum gross area in tension perpendicular to the line of force
   $= 32 \times 20 = 640\,\text{mm}^2$

Atn = minimum net area in tension perpendicular to the line of force

$$= 640 - 2 \times 0.5 \times 22 \times 20 = 200 \, \text{mm}^2$$

$$\text{Tdb1} = \left[10720 \times 250 / (1.732 \times 1.1) + 0.9 \times 200 \times 410 / 1.25 \right] / 1000 = 1466 \, \text{kN}$$

$$\text{Tdb2} = \left[0.9 \times 7200 \times 410 / (1.732 \times 1.25) + 640 \times 250 / 1.1 \right] / 1000 = 1373 \, \text{kN}$$

Tdb = 1373 kN; fat = 0.69 Tdb/Ag = **208** MPa
fat for design = minimum of fat in a), b), and c) above = **150** MPa
Flange force, $F1 = 150 \times 210 \times 20.8/1000 = $ **655** kN

Design of connection bolts (working stress) (IS 800:2007 – 11.6.1)
Let us use,

M20	HT bolts 8.8 grade, bearing type
Bolt diameter	d = 20 mm
$\gamma_{mb} = 1.25$	Material Safety factor.
Fub = 830 MPa	Ultimate tensile stress of bolt.
Fu = 410 MPa	Ultimate tensile stress of plate.
$n_n = 0$	No. of shear planes with threads intercepting the shear plane.
$n_s = 1$	No. of shear planes without threads intercepting the shear plane.
Asb = 314 mm²	Nominal plain shank area of the bolt.
Anb = 245 mm²	Net shear area of the bolt at thread.
$e_1 = 32$ mm	End distance along bearing direction.
$d_0 = 22$ mm	Diameter of the hole.
p = 50mm	Pitch distance along bearing direction (2.5 d)
kb = 0.485	[kb = smaller of e/3d0, (p/3d0 – 0.25), fub/fu, 1].

Shear capacity:
Vnsb = (Fub/ √ 3) × (n_n. Anb + n_s. Asb) (IS:800 – 2007 – 10.3.3)

$$\text{Vnsb} = \left[(830 / 1.732) \times (1 \times 314) \right] / 1000 = 150 \, \text{kN}$$

Bolt value, Vsb = 0.6 Vnsb = 0.6 × 150 = **90** kN (single shear)
[Long joint factor: (IS:800 – 2007 – 10.3.3.1)
Lj = distance between first and last bolts = 150 mm [(4-1) × 50]
Allowable limit without any reduction of stress = 15d = 300 mm. Hence,
Not applicable.]

Bearing capacity:
Bearing Plate thickness = **20** mm [smaller of flange and cover plate thick-
ness].
Apb = 20 × 20 = 400 mm² bearing area on plate (d × t)
Vnpb = nominal bearing strength of a bolt = 2.5 kb .d .t. Fu

(IS:800 – 2007 – 10.3.4)

$$\text{Vpnb} = 2.5 \times 0.485 \times 20 \times 20 \times 410 / 1000 = 199 \text{ kN}$$

Bolt value, Vsb = 0.6 Vpnb = 0.6 × 199 = **119** kN
Design force, **F = 655 kN** (axial)
Now, Shear Capacity of a bolt = 90 kN (Single shear)
Bearing Capacity of a bolt = 119 kN
Bolt value or Design strength, Vsb = **90** kN
Number of bolts (N1) = 8 (2 × 4)

$$\text{Total Capacity} = 8 \times 90 = \textbf{720} \text{ kN} \quad > F = \textbf{655 kN; Safe.}$$

f) Sizing of Web Cover Plate and Connection Bolts:

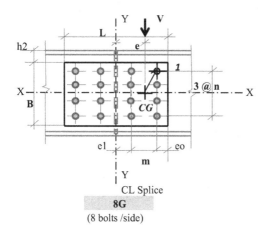
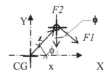

$F_1 = Ve \cdot z / \sum (x^2 + y^2) \quad I = \sum (x^2 + y^2)$
$F2 = V / n_o$
$F = \sqrt{(F_1{}^2 + F_2{}^2 + 2 F_1 F_2 \cos \phi)}$

FIGURE 4.31 Arrangement of bolts in web splice.

Beam section spliced: **ISMB 600**
Web cover plate -1; No. of plates = 2
wide, Bw = 275 mm; depth, dw = 500 mm; thick, t_{wc} = 8 mm
Bolt
Group = **8G**; Nos / side, n_0 = 8; Bolt size = M20; Bolt dia., d = 20 mm
e_0 = 32 mm; e_1 = 33.5 mm; m = 60 mm

$$n = (500 - 2 \times 32) / (4 - 1) = 145 \text{ mm}$$

Total no. of bolts at web splice = 2 × 8 = 16; Properties of group:

$$Ixx = 218418 \text{ mm}^2$$

$$z = 220 \text{ mm}$$

$$\cos \phi = 0.136$$

Web area of spliced beam, $Aw = 600 \times 12 = 7200 \text{ mm}^2$

Sectional area of cover plates $= 2 \times 500 \times 8 = 8000 \text{ mm}^2 > \textbf{1.05 Aw; Okay.}$

To determine the maximum load on the bolts connecting web:

There are two groups of bolts symmetrically placed on either side of splice center line. The load (V) and moment (V × e) will be resisted by all the bolts in each group seperately. Therefore, the maximum load will be on the furthest bolt from the center of gravity of the group.

Web shear, $V = 50\%$ (0.4 fy Aw)/1000 $= 0.5 \times 0.4 \times 250 \times 600 \times 12/1000$
$= \textbf{360}$ kN

Eccentricity, $e = e_1 + 0.5 \times m = 33.5 + 0.5 \times 60 = 63.5 \text{ mm}$
$F_1 = \text{Ve.} \ z/ \Sigma \ (x^2 + y^2); \ I = \Sigma \ (x^2 + y^2)$

$$F_2 = V / n_o$$

$$\textbf{Fb} = \sqrt{\left(\textbf{F}_1{}^2 + \textbf{F}_2{}^2 + 2 \ \textbf{F}_1\textbf{F}_2 \cos\phi \right)}$$

$$F_1 = 360 \times 63.5 \times 220 / 218418 = 23 \text{ kN}$$

$$F_2 = 360 / 8 = 45 \text{ kN}$$

$$F = \sqrt{\left(23^2 + 45^2 + 2 \times 23 \times 45 \times 0.136 \right)} = \textbf{51} \text{ kN}$$

Bolt value calculation
  Let us use,

M20	HT bolts 8.8 grade, bearing type
Bolt diameter	d = 20 mm
$\gamma_{mb} = 1.25$	Material Safety factor.
Fub = 830 MPa	Ultimate tensile stress of bolt.
Fu = 410 MPa	Ultimate tensile stress of plate.
$n_n = 0$	No. of shear planes with threads intercepting the shear plane.
$n_s = 2$	No. of shear planes without threads intercepting the shear plane.
Asb = 314 mm²	Nominal plain shank area of the bolt.
Anb = 245 mm²	Net shear area of the bolt at thread.
$e_1 = 32$ mm	End distance along bearing direction.
$d_0 = 22$ mm	Diameter of the hole.
p = **60** mm	Pitch distance along bearing direction (= m).
kb = 0.485	[kb = smaller of e/3d0, (p/3d0 – 0.25), fub/fu, 1].

Shear capacity:
Vnsb $= (Fub/ \sqrt{3}) \times (n_n.\ Anb + n_s.\ Asb)$ (IS:800 – 2007 – 10.3.3)

$$\text{Vnsb} = \left[\left(830 / 1.732 \right) \times \left(2 \times 314 \right) \right] / 1000 = 301 \text{ kN}$$

Bolt value, Vsb = 0.6 Vnsb $= 0.6 \times 301 = \textbf{181}$ kN (double shear)

Bearing capacity:

Bearing Plate thickness = **12** mm [smaller of web and sum of cover plate thickness].

Bearing area, Apb = 20 × 12 = 240 mm² (d × t)

Vnpb = nominal bearing strength of a bolt = 2.5 kb .d .t. Fu (IS:800 – 2007 – 10.3.4)

$$\text{Vpnb} = 2.5 \times 0.485 \times 20 \times 12 \times 410 / 1000 = 119 \,\text{kN}$$

Bolt value, Vsb = 0.6 Vpnb = 0.6 × 119 = **71** kN

Design force, **F = 51 kN**(max. load on the furthest bolt from the center of gravity of the group.)

Now, Shear Capacity of a bolt = 181 kN (double shear)

Bearing Capacity of a bolt = 71 kN

Bolt value or Design strength, Vsb = **71** kN per bolt

Number of bolts at furthest corner (*mkd.1*) = **1**

$$\text{Total Capacity} = 1 \times 71 = \textbf{71}\,\text{kN} \quad > \textbf{F = 51 kN; Safe}.$$

g) **Summary of Calculation**

Spliced member: **ISMB 600**

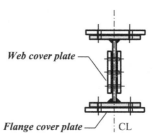

FIGURE 4.32   Sectional view at splice.

Design force:

Web shear, V = 360 kN [50% web shear considered at splice location]

Moment, M = 407 kNm [Full strength Moment connection]

Connection Materials:

1. Flange Cover plates – 2 nos 230 × 20 × 600 long
2. Web Cover plates – 2 nos 275 × 8 × 500 long
3. Bolts: M20 HT bolts grade 8.8 (bearing type connection)

$$\text{At Flanges} - 2 \times 8 = 16 \,\text{at top flange and}$$
$$2 \times 8 = 16 \,\text{at bottom flange.}$$
$$\text{At Web} - \quad 2 \times 8 = 16$$

Total numbers of bolts = 48

4.6.2 BEAMS AND GIRDERS WITH FRICTION GRIP BOLT – IS 800 (LIMIT STATE)

a) Sketch
Please see Figure 4.20
b) Materials
Same as in Table 4.6
c) Member Descriptions:
Spliced member: **ISMB 600 Type – 1**
Flange plates: Web Plate:
wide, bf = 210 mm; depth, d = 600 mm
thickness, tf = 20.8 mm; thickness, tw = 12 mm
Connection Bolt: **M20**; **HT Bolt 8.8 grade – Friction Grip bolt**
d) Design Parameters:
1. Flange force, F: Full tension capacity
2. Web shear, V: 50% of web shear capacity
3. Sum area of cover plates is 5% more than the area of section spliced.
4. All bolts are 8.8 grade HT bolts Friction grip.
e) Sizing of Flange Cover Plate and Connection Bolt:
Let us consider following dimensions for flange cover plates:
Flange cover plate –1:
wide, Bf1 = 230 mm; thick, Tf = 20 mm; length, Lf = 750 mm
No. of plate per flange = 1 per flange
Flange cover plate 2:
wide, Bf_2 = 90 mm; thick, Tf = 20 mm; length, Lf = 750 mm
No. of plate per flange = 2 per flange
Cross sectional area of girder flange, Af = 210 × 20.8 = 4368 mm^2
Sectional area of cover plate, A required = Af × 1.05 = 4368 × 1.05 = 4586 mm^2
Total sectional area of cover plates,

$$A \text{ provided} = 1 \times 230 \times 20 + 2 \times 90 \times 20 = 8200 \text{ mm}^2 > A \text{ required; Okay.}$$

Flange connection bolts, N1:
Refer Figure 4.30: Plan view of top cover plate 1 in 4.6.1.
Bolt: M24 HT 8.8 grade – Friction Grip
Bolt dia., d = 24 mm; Hole dia., do = 27 mm; Edge dist., e_0 = 44 mm
No. of rows = **2**; Nos of Bolt / row = **5**

$$Lw = (750/2) - 44 = 331 \text{ mm}$$

Spacing of bolt, $s = (375 - 44 \times 2)/(5-1) = 72$ mm > 2.5 d; Okay.
Tension in cover plate:
The actual tensile stress on the cover plate shall be smaller of the following:
a) Yielding of gross section, fat = fy/γ_{mo} (IS 800:2007 – 6.2)
b) Rupture of the net section, fat = 0.9 fu/γ_{m1}
c) Block shear resistance, fat = Tdb/Ag

a) by yielding:

$$fat = 250 / 1.1 = \mathbf{227} \, MPa$$

b) by rapture: (IS 800:2007 – 6.3.1)

$$An = 230 \times 20 - 2 \times 27 = 4546 \, mm^2$$

$$Tdn = 0.9 \, An \, fu / \gamma_{m1} = 0.9 \times 4546 \times 410 / 1.25 / 1000 = 1342 \, kN$$

$$fat = 1342000 / 4546 = \mathbf{295} \, MPa$$

c) by block shear:
   Cover plate 1: Lw = 331 mm; $b = e_0 = 44$ mm
   t = 20 mm
   Tb = Smaller of Tdb1 or Tdb2 (IS 800:2007 – 6.4.1; 6.4.2)

$$\boldsymbol{Tdb1 = \left[Avg.fy / \left(\sqrt{3} \, \gamma_{mo} \right) + 0.9 \, Atn \, fu / \gamma_{m1} \right]}$$

$$\boldsymbol{Tdb2 = \left(0.9 A_{vn} fu / \left(\sqrt{3} \, \gamma_{m1} \right) + A_{tg} fy / \gamma_{mo} \right)}$$

Avg = gross area in shear along bolt line $= 2 \times 331 \times 20 = 13240 \, mm^2$

Avn = net area in shear along bolt line $= 13240 - 2 \times 5 \times 27 \times 20 = 7840 \, mm^2$

Atg = minimum gross area in tension perpendicular to the line of force
$= 44 \times 20 = 880 \, mm^2$

Atn = minimum net area in tension perpendicular to the line of force

$$= 880 - 2 \times 0.5 \times 27 \times 20 = 340 \, mm^2$$

$$Tdb1 = \left[13240 \times 250 / \left(1.732 \times 1.1 \right) + 0.9 \times 340 \times 410 / 1.25 \right] / 1000 = 1838 kN$$

$$Tdb2 = \left[0.9 \times 7840 \times 410 / \left(1.732 \times 1.25 \right) + 880 \times 250 / 1.1 \right] / 1000 = 1536 \, kN$$

Tdb = 1536 kN; fat = Tdb/Ag = **338** MPa
   fat for design = minimum of fat in a), b), and c) above = **227** MPa
   Flange force, **F1** = 227 × 210 × 20.8/1000 = **992** kN
   Design of connection bolts (Friction grip, Limit state) (IS 800:2007 –
   10.4; 10.3.4)

Let us use,

M20	HT bolts 8.8 grade, bearing type
Bolt diameter	d = 24 mm
$\gamma_{mb} = 1.25$	Material Safety factor.
Fub = 830 MPa	Ultimate tensile stress of bolt.
Fu = 410 MPa	Ultimate tensile stress of plate.
$n_n = 0$	No. of shear planes with threads intercepting the shear plane.
$n_s = 2$	No. of shear planes without threads intercepting the shear plane.
Asb = 453 mm²	Nominal plain shank area of the bolt.
Anb = 353 mm²	Net shear area of the bolt at thread.
$e_1 = 44$ mm	End distance along bearing direction.
$d_0 = 27$ mm	Diameter of the hole.
p = 72 mm	Pitch distance along bearing direction (3 d).
kb = 0.543	[kb = smaller of e/3d0, (p/3d0 – 0.25), fub/fu, 1].

Slip-critical connections shall be designed to prevent slip and for the limit states of bearing-type connections.

(i) Slip resistance for bearing, Vdpb = Vnpb/γ_{mb} (IS 800:2007 – 10.3.4)
  Where, d = 24 mm; do = 27 mm; t = 20.8 mm

$$V\,pnb = 2.5\,kb\,d\,t\,Fu = 2.5 \times 0.543 \times 24 \times 20.8 \times 410\,/\,1000 = 267\,kN$$

$$Vsb = 267\,/\,1.25 = \mathbf{214}\,kN$$

(ii) Slip resistance for pretension, **Vsf = Vnsf/γ_{mf}** (IS 800:2007 – 10.4)
  Where, $\gamma_{mf} = 1.10$ at service load

$$\mathbf{Vnsf = \mu f.n_e.Kh.F_0}$$

where,
  $\mathbf{\mu f}$ = 0.33 slip factor (IS 800:2007 – Table 20)
  $\mathbf{n_e}$ = 2 no. of slip surface
  Kh = 1 factor for clearance hole
  $\mathbf{F_0}$ = 205 kN (Anb. 0.7fub = 353 × 0.7 × 830 / 1000 = 205 kN)

$$\mathbf{Vsf = \left(0.33 \times 2 \times 1 \times 205\right)/1.25 = 108\,kN}$$

Bolt value = **108** kN (lower of Vsf and Vsb)
Flange force, F = 992 kN Number of bolts (N1) = 2 × 5 = 10

$$\text{Total Capacity} = 10 \times 108 = \mathbf{1080\,kN} > \mathbf{F = 992\,kN; Safe.}$$

f) Sizing of Web Cover Plate and Connection Bolts:

Please see Figure 4.31 in example 4.6.1

Beam section spliced: **ISMB 600**

Web cover plate –1; No. of plates = 2

wide, Bw = 275 mm; depth, dw = 500 mm; thick,t_{wc} = 8 mm

Bolt

Group = **8G**; Nos / side, n_0 = **8**; Bolt size = **M24**; Bolt dia., d = 24 mm

e_0 = 44 mm; e_1 = 45.5 mm; m = 72 mm

$$n = (500 - 2 \times 44) / (4 - 1) = 137 \, mm$$

Total no. of bolts at web splice = 2 × 8 = 16

Properties of group: Ixx = 198972 mm^2; z = 209 mm;   cos ϕ = 0.172

Web area of spliced beam, Aw = 600 × 12 = 7200 mm^2

Sectional area of cover plates = $2 \times 500 \times 8 = 8000 \, mm^2$ > **1.05 Aw; Okay.**

To determine the maximum load on the bolts connecting web:

Please see Figure 4.31 in example 4.6.1.

There are two groups of bolts symmetrically placed on either side of splice center line. The load (V) and moment (V × e) will be resisted by all the bolts in each group separately. Therefore, the maximum load will be on the furthest bolt from the center of gravity of group.

Web shear, V = 50 % [(fy/ $\sqrt{3}$). Aw/γ_{m0}]/1000

$$= \left[0.5 (250 / 1.732) \times 7200 / 1.1 \right] / 1000 = \textbf{472} \, \textbf{kN}$$

Eccentricity, e = e_1 + 0.5 × m = 45.5 + 0.5 × 72 = 81.5 mm

$$F_1 = Ve.z / \Sigma (x^2 + y^2) \qquad\qquad I = \Sigma (x^2 + y^2)$$

$$F_2 = V / n_o$$

$$\textbf{Fb} = \sqrt{\left(\textbf{F}_1{}^2 + \textbf{F}_2{}^2 + \textbf{2} \, \textbf{F}_1 \textbf{F}_2 \cos \phi \right)}$$

$$F_1 = 472 \times 81.5 \times 209 / 198972 = 40 \, kN$$

$$F_2 = 472 / 8 = 59 \, kN$$

$$\textbf{F} = \sqrt{\left(40^2 + 59^2 + 2 \times 40 \times 59 \times 0.172 \right)} = \textbf{71} \, \textbf{kN}$$

Bolt value calculation (Friction Grip; Limit state) (IS 800:2007 – 10.4; 10.3.4)

Let us use,

M24	HT bolts 8.8 grade, Friction Grip
Bolt diameter	d = 24mm
$\gamma_{mb} = 1.25$	Material Safety factor.
Fub = 830 MPa	Ultimate tensile stress of bolt.
Fu = 410 MPa	Ultimate tensile stress of plate.
$n_n = 0$	No. of shear planes with threads intercepting the shear plane.
$n_s = 2$	No. of shear planes without threads intercepting the shear plane.
Asb = 453 mm^2	Nominal plain shank area of the bolt.
Anb = 353 mm^2	Net shear area of the bolt at thread.
$e_1 = 44$ mm	End distance along bearing direction.
$d_0 = 27$ mm	Diameter of the hole.
p = 72 mm	Pitch distance along bearing direction.
kb = 0.543	[kb = smaller of e/3d0, (p/3d0 – 0.25), fub/fu, 1].

Slip-critical connections shall be designed to prevent slip and for the limit states of bearing-type connections.

1. Slip resistance for bearing, Vdpb = Vnpb/γ_{mb} (IS 800:2007 – 10.3.4)
   Where, d = 24 mm; do = 27 mm; t = 12 mm

$$V\,pnb = 2.5\,kb\,d\,t\,Fu = 2.5 \times 0.543 \times 24 \times 12 \times 410 / 1000 = 154\ kN$$

$$Vsb = 154 / 1.25 = \mathbf{123}\,kN$$

2. Slip resistance for pretension, **Vsf = Vnsf/γ_{mf}** (IS 800:2007 – 10.4)
   where, $\gamma_{mf} = 1.10$ at service load

$$\mathbf{Vnsf = \mu f.n_e.Kh.F_0}$$

where,
   μf = 0.33 slip factor (IS 800:2007 – Table 20)
   n_e = 2 no. of slip surface
   Kh = 1 factor for clearance hole
   F_0 = 205 kN (Anb. 0.7 fub = 353 × 0.7 × 830 / 1000 = 205 kN)

$$\mathbf{Vsf} = \left(0.33 \times 2 \times 1 \times 205\right) / 1.25 = \mathbf{108\,kN}$$

Bolt value = **108** kN (lower of Vsf and Vsb)
Design force, **F = 71 kN** (max. load on the furthest bolt from the center of gravity of the group)
< Bolt Value =108 kN; hence, Safe.

g) Summary of Calculation

Spliced member: **ISMB 600**

Design force:

Web shear, V = 472 kN [50% web shear considered at splice location]

Moment, M = 616 kNm [Full strength Moment connection]

Connection Materials:

1. Flange Cover plates 1: – 1 nos 230 × 20 × 750 long
2. Flange Cover plates 2: – 2 nos 90 × 20 × 750 long
3. Web Cover plates: – 2 nos 275 × 8 × 500 long
4. Bolts: M24 HT bolts grade 8.8 (Friction grip connection)
5. At Flanges – 2 × 10 = 20 at top flange and
   2 × 10 = 20 at bottom flange.
6. At Web – 2 × 8 = 16
   Total no. of bolts = 56

4.6.3   CHANNEL SECTIONS WITH BEARING TYPE BOLT – IS 800 (WORKING STRESS)

a) Sketch

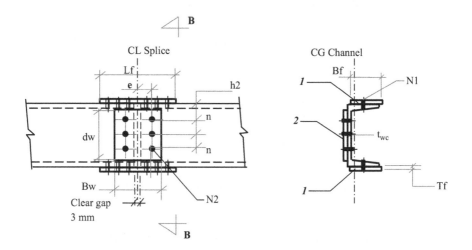

Elevation   Section B-B

FIGURE 4.33   Channel member splice.

b) Materials

See Table 4.6

c) Member Descriptions
Spliced member: **ISMC 250**

Flange:	*Web*:
wide, bf = 80 mm	depth, d = 250 mm
thickness, tf = 14.1 mm	thickness, tw = 7.1 mm

Bolt: **M20**; HT bolt 8.8 grade

d) Design Parameters:
1. Flange force, F: Full tension capacity = 0.6 fy. Af; where, Af = flange area
2. Web shear, V: 50% of web shear capacity (= 0.4 fy Aw); where, Aw= web area
3. Sum area of cover plates is 5% more than the area of section spliced.
4. All bolts are 8.8 grade HT bolts bearing type.

e) Sizing of Flange Cover Plate and Connection Bolt:
Let us consider following dimensions for flange cover plates:
Flange cover plate -1:
wide, Bf1 = 80 mm; thick, Tf =16 mm; length, Lf = 500 mm
No. of plates per flange = 1
Cross sectional area of girder flange, Af = 80 × 14.1 = 1184 mm²
Sectional area of cover plate, A required = Af × 1.05 = 1128 × 1.05 = 1184 mm²
Total sectional area of cover plates, A provided = 1 × 80 × 16 = 1280 mm² > A required; Okay.

Bolt connecting flanges and cover plate, N1:

Bolt: M20; HT 8.8 grade, bearing type
Bolt dia., d = 20 mm
Hole dia., do = 22 mm
Edge dist., e_0 = 32 mm
No. of rows = **1**
Nos of Bolt / row = **3**

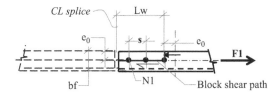

FIGURE 4.34   Plan view of channel at splice.

$$\mathbf{Lw} = Lf / 2 - e_0 = 0.5 \times 500 - 32 = 218 \, mm$$

$$\text{Spacing of bolt}, s = (0.5 \times 500 - 32 \times 2) / (3 - 1) = 93 \, mm > 2.5 \, d; \text{Okay.}$$

Tension in cover plate:

The actual tensile stress on the cover plate shall be smaller of the following:

1. Yielding of gross section, fat = 0.6 fy (IS 800:2007 – 11.2.1)
2. Rupture of the net section, fat = 0.69 Tdn/Ag
3. Block shear resistance, fat = 0.69 Tdb/Ag

a) by yielding:

$$\text{fat} = 0.6 \times 250 = \textbf{150} \text{ MPa}$$

b) by rapture: (IS 800:2007 – 6.3.1)

$$\text{An} = 80 \times 16 - 22 \times 16 = 928 \text{ mm}^2$$

$$\text{Tdn} = 0.9 \text{ An fu} / \gamma_{ml} = 0.9 \times 928 \times 410 / 1.25 / 1000 = 274 \text{ kN}$$

$$\text{fat} = 0.69 \times 274000 / 928 = \textbf{204} \text{ MPa}$$

c) by block shear:

Cover plate: Lw = 218 mm; $b = e_0 = 32$ mm

$$t = 16 \text{ mm}$$

Tb = Smaller of Tdb1 or Tdb2 (IS 800:2007 – 6.4.1; 6.4.2)

$$\textbf{\textit{Tdb1}} = \left[\textbf{\textit{Avg.fy}} / \left(\sqrt{3} \, \gamma_{mo} \right) + \textbf{\textit{0.9 Atn fu}} / \gamma_{ml} \right]$$

$$\textbf{\textit{Tdb2}} = \left(\textbf{\textit{0.9A}}_{vn} \textbf{\textit{fu}} / \left(\sqrt{3} \, \gamma_{ml} \right) + A_{tg} \textit{fy} / \gamma_{mo} \right)$$

Avg = gross area in shear along bolt line $= 1 \times 218 \times 16 = 3488 \text{ mm}^2$

Avn = net area in shear along bolt line $= 3488 - 1 \times 3 \times 22 \times 16 = 2432 \text{ mm}^2$

Atg = minimum gross area in tension perpendicular to the line of force
$= 32 \times 16 = 512 \text{ mm}^2$

Atn = minimum net area in tension perpendicular to the line of force

$$= 512 - 1 \times 0.5 \times 22 \times 16 = 336 \text{ mm}^2$$

$$\text{Tdb1} = \left[3488 \times 250 / (1.732 \times 1.1) + 0.9 \times 336 \times 410 / 1.25 \right] / 1000 = 557 \text{ kN}$$

$$\text{Tdb2} = \left[0.9 \times 2432 \times 410 / (1.732 \times 1.25) + 512 \times 250 / 1.1 \right] / 1000 = 531 \text{ kN}$$

Tdb = 531 kN; fat = 0.69 Tdb/Ag = **395** MPa (IS 814:2004 – 7.5.1.2)
    fat for design = minimum of fat in a), b), and c) above = **150** MPa
Flange force, $F1 = 150 \times 80 \times 14.1/1000 =$ **169** kN
Design of connection bolts (working stress) (IS 800:2007 – 11.6.1)
Let us use,

M20	HT bolts 8.8 grade, bearing type
Bolt diameter	d = 20 mm
$\gamma_{mb} = 1.25$	Material Safety factor.
Fub = 830 MPa	Ultimate tensile stress of bolt.
Fu = 410 MPa	Ultimate tensile stress of plate.
$n_n = 0$	No. of shear planes with threads intercepting the shear plane.
$n_s = 1$	No. of shear planes without threads intercepting the shear plane.
Asb = 314 mm²	Nominal plain shank area of the bolt.
Anb = 245 mm²	Net shear area of the bolt at thread.
$e_1 = 32$ mm	End distance along bearing direction.
$d_0 = 22$ mm	Diameter of the hole.
p = 50 mm	Pitch distance along bearing direction (2.5 d).
kb = 0.485	[kb = smaller of e/3d0, (p/3d0 – 0.25), fub/fu, 1].

Shear capacity:
    Vnsb = (Fub/ $\sqrt{3}$) $\times$ (n_n. Anb + n_s. Asb) (IS: 800 – 2007 – 10.3.3)

$$V\text{nsb} = \left[(830/1.732) \times (1 \times 314)\right]/1000 = 150 \text{ kN}$$

Bolt value, Vsb = 0.6 Vnsb = 0.6 × 150 = **90** kN (single shear)

Bearing capacity:
Bearing Plate thickness = **14.1** mm [smaller of flange and cover plate thickness].
    Apb = 20 × 14.1 = 282 mm² bearing area on plate (d × t)
    Vnpb = nominal bearing strength of a bolt = 2.5 kb .d .t. Fu
    (IS: 800 – 2007 – 10.3.4)

$$V\text{pnb} = 2.5 \times 0.485 \times 20 \times 14.1 \times 410/1000 = 140 \text{ kN}$$

Bolt value, Vsb = 0.6 Vpnb = 0.6 × 140 = **84** kN
Design force, **F = 169 kN** (axial)
    Now, Shear Capacity of a bolt = 90 kN (Single shear)
    Bearing Capacity of a bolt = 84 kN
    Bolt value or Design strength, Vsb = **84** kN
    Number of bolts (N1) = 3

    Total Capacity = 3×84 = **252** kN > **F = 169 kN**; **Safe**.

f) Dimensioning Web Cover Plate and Connection Bolt:

Channel section spliced: **ISMC 250**

Web cover plate -1; No. of plates = 1

wide, Bw = 200 mm; depth, dw = 200 mm; thick, t_{wc} = 10 mm

Bolt

Nos / side, n_0 = 3; Bolt size = M20; Bolt dia., d = 20 mm; e_0 = 32 mm

Web area of spliced beam, Aw = 250 × 7.1 = 1775 mm²

Sectional area of cover plate = 1 × 200 × 10 = 2000 mm²>**A required; Okay**.

Web shear, V = 50 % (0.4 fy Aw)/1000 = 0.5 × 0.4 × 250 × (250 × 7.1)/1000 = **89** kN

Moment on group = V.e $\left(e = \text{eccentricity}\right)$ = 88.5 × 33.5 / 1000 = 2.97 kNm

$$F1 = V.e / 2\,n$$

$$F2 = V / n_o$$

$$\mathbf{F = \sqrt{\left(F_1^2 + F_2^2 + 2\,F_1\,F_2 \cos\phi\right)}}$$

$$z = 98.85\,\text{mm}$$

$$\cos\phi = 0.3389$$

$$n = 93\,\text{mm}$$

$$e = 33.5\,\text{mm}$$

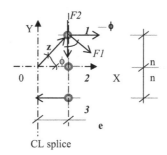

FIGURE 4.35 Force diagram.

$$F_1 = 2970 / \left(2 \times 93\right) = 15.97\,\text{kN}$$

$$F_2 = 88.75 / 3 = 29.58\,\text{kN}$$

$$\mathbf{F = 38.08\,kN}$$

Bolt value calculation
    Let us use,

M20	HT bolts 8.8 grade, bearing type
Bolt diameter	d = 20 mm
γ_{mb} = 1.25	Material Safety factor.
Fub = 830 MPa	Ultimate tensile stress of bolt.
Fu = 410 MPa	Ultimate tensile stress of plate.
n_n = 0	No. of shear planes with threads intercepting the shear plane.
n_s = 1	No. of shear planes without threads intercepting the shear plane.
Asb = 314 mm^2	Nominal plain shank area of the bolt.
Anb = 245 mm^2	Net shear area of the bolt at thread.
e_1 = 32 mm	End distance along bearing direction.
d_0 = 22 mm	Diameter of the hole.
p = **50** mm	Pitch distance along bearing direction.
kb = 0.485	[kb = smaller of e/3d0, (p/3d0 – 0.25), fub/fu, 1].

Shear capacity:

$$\text{Vnsb} = \left[(830 / 1.732) \times (1 \times 314) \right] / 1000 = 150 \, \text{kN}$$

Vnsb = (Fub/ $\sqrt{3}$) × (n_n. Anb + n_s. Asb) (IS: 800 – 2007 – 10.3.3)
Bolt value, Vsb = 0.6 Vnsb = 0.6 × 150 = **90** kN (single shear)

Bearing capacity:
Bearing Plate thickness = **7.1** mm [smaller of web and sum of cover plates thickness].
    Bearing area, Apb = 20 × 7.1 = 142 mm^2 ($d \times t$)
    Vnpb = nominal bearing strength of a bolt = 2.5 kb .d .t. Fu
    (IS: 800 – 2007 – 10.3.4)

$$\text{Vpnb} = 2.5 \times 0.485 \times 20 \times 7.1 \times 410 / 1000 = 71 \, \text{kN}$$

    Bolt value, Vsb = 0.6 Vpnb = 0.6 × 71 = **43** kN

Block shear capacity:
Web cover plate:
Width, Bw = 200 mm; depth, dw = 200 mm; thick, t_{wc} = 10 mm
Lw = 168 mm; $b = e_0$ = 32 mm; t = 00 mm
Tb = Smaller of Tdb1 or Tdb2 (IS 800:2007 – 6.4.1; 6.4.2)

$$\boldsymbol{Tdb1 = \left[Avg.fy / \left(\sqrt{3} \, \gamma_{mo} \right) + 0.9 \, Atn \, fu / \gamma_{m1} \right]}$$

$$\boldsymbol{Tdb2 = \left(0.9 A_{vn} fu / \left(\sqrt{3} \, \gamma_{m1} \right) + A_{tg} fy / \gamma_{mo} \right)}$$

Avg = gross area in shear along bolt line $= 1 \times 168 \times 10 = 1680 \, \text{mm}^2$

Avn = net area in shear along bolt line $= 1680 - 1 \times 3 \times 22 \times 10 = 1020 \, \text{mm}^2$

Atg = minimum gross area in tension perpendicular to the line of force
$= 32 \times 10 = 320 \, \text{mm}^2$

Atn = minimum net area in tension perpendicular to the line of force $= 210 \, \text{mm}^2$

$$\text{Tdb1} = \left[1680 \times 250 / \left(1.732 \times 1.1 \right) + 0.9 \times 210 \times 410 / 1.25 \right] / 1000 = 282 \, \text{kN}$$

$$\text{Tdb2} = \left[0.9 \times 1020 \times 410 / \left(1.732 \times 1.25 \right) + 320 \times 250 / 1.1 \right] / 1000 = 247 \, \text{kN}$$

Tdb = 247 kN; Vsb = 0.69 Tdb = **170** MPa; (IS 800:2007 – 11.2.1)
Bolt value = Smaller of a), b), and c) = **43 kN > F = 38 kN; Safe.**
No. of Bolts = 3
Total capacity of bolts = 3 × 43 = **129** kN
Design web Shear, V = **89** kN < 129; hence, Safe.

4.6.4 BEAM AND GIRDER WITH SLIP CRITICAL BOLT – AISC (ASD)

a) **Sketch**
Refer Figure 4.20 in example 4.6.
b) **Material**
Refer Table 4.7.
c) **Member Descriptions**
Spliced member: **W24 × 84; Type – 1**

Flange plates:	*Web Plate*:
wide, bf = 9.02 inch	depth, d = 24.1 inch
thickness, tf = 0.77 inch	thickness, tw = 0.47 inch

Bolt: **M20; A490M; Slip-critical connection (ASD)**
d) **Design Parameters**
1. Flange force, F: Full tension capacity = 0.6 fy Af, where, Af = flange area
2. Web shear, V: 50% of web shear capacity (0.6 fy Aw/Ω)
3. Sum area of cover plates is 5% more than the area of section spliced.
4. All bolts are High tensile bolt Slip-critical connection.
e) **Flange Cover Plate and Bolt**
Let us consider following dimensions for flange cover plates:
Flange cover plate –1:
wide, Bf_1 = 8 inch; thick, Tf = 0.625inch; length, Lf = 30 inch
No. of plate per flange = 1 per flange

Flange cover plate 2:

wide, Bf_2 = 2.77 inch; thick, Tf = 0.625 inch; length, Lf = 30 inch

No. of plate per flange = 2 per flange

Cross sectional area of girder flange, Af = 9.02 × 0.77 = 6.95 inch^2

Sectional area of cover plate, *A* required = Af × 1.05 = 6.95 × 1.05 = 7.3 inch^2

Total sectional area of cover plates,

A provided = 1 × 8 × 0.625 + 2 × 2.77 × 0.625 = 8.46 inch^2 > A required; Okay.

Flange force in cover plate:

Refer Figure 4.30 Plan view of cover plate (but no. of bolts /row will be as given below).

Bolt: **M20**; A490M

Bolt dia., d = 20 mm; 1.00 inch

Hole dia., do = 22 mm; 0.87 inch

Edge dist., e_0 = 32 mm; 1.26 inch

No. of rows = **2**; Nos of Bolt / row = **5**

Gross area, Ag = 8 × 0.625 = 5.00 inch^2

Net area, An = 5 − 2 × 0.87 × 0.625 = 3.91 inch^2

$$\textbf{Lw} = 0.5 \times 30 - 1.26 = \textbf{13.74 inch}$$

Spacing of bolt, s $= (15 - 1.26 \times 2)/(5 - 1) = 3.12$ inch   s > 2.5 d; OK.

Cover plate 1:

Tensile strength:

1. Tensile yielding, Pn1 = FyAg/Ω = 36 × 5/1.67 = 108 kips
   Ω = 1.67ASD φ = 0.9 LRFD
2. Tensile rupture, Pn2 = FuAe/Ω = Fu An. U/Ω = 65 × 3.91 × 1/2 = 127 kips
   Ω = 2 ASD φ = 0.75 LRFD Ae = An.U
   U = shear lag factor as in Table D3.1 U = 1 with Case 4
   Allowable Tension, T = 108 kips

To determine block shear resistance, V = Rn/Ω:

Cover plate 1

L = Lw = 13.74 inch; 2L = 27.48 inch

t = 0.63 inch; $b = e_0$ = 1.26 inch; 2b = 2.52 inch

$$Rn = 0.6\,Fu\,Anv + Ubs\,Fu\,Ant \quad <= 0.6\,Fy\,Agv + Ubs\,Fu\,Ant$$

Ω = 2 ASD; φ = 0.75 LRFD

Where,

$$Agv = \text{gross area subjet to shear yielding} \left[2\,L.t \right]$$
$$= 27.48 \times 0.625 = 17.18\,\text{inch}^2$$

Anv = net area subject to shear rupture $\left(\text{Agv- }\Sigma\text{hole area}\right)$

$$= 17.18 - 2\times5\times0.87\times0.625 = 11.74\,\text{inch}^2$$

Ant = net area subject to tension $\left(\text{b.t}\right) = 2.52\times0.625 = 1.58\,\text{inch}^2$

Ubs = 1, where tension stress is uniform (0.5 where tension stress is non-uniform)
Now,

$$0.6\,\text{Fu Anv} + \text{Ubs Fu Ant} = 0.6\times65\times11.74 + 1\times65\times1.58 = 561\,\text{kips}$$

$$0.6\,\text{Fy Agv} + \text{Ubs Fu Ant} = 0.6\times36\times17.18 + 1\times65\times1.58 = 474\,\text{kips}$$

Rn = 474 kips $V = \text{Rn}/\Omega = 474/2 = 237$ kips
Block shear strength, $V = \textbf{237}$ kips < Tension.
Design value should be smaller of tension and block shear strength.
Maximum tension in top cover plate = **108 kips**
Cover Plate 2
$\text{Pn}_1 = 37.32$ kips; $\text{Pn}_2 = 39$ kips; T = 37 kips
Rn = Not applicable for single row of bolts
Maximum tension in bottom cover plates = $2 \times 37 = 74$ kips
Total resistance of cover plates = $108 + 74 = 182$ kips
Flange force of spliced member:
wide, bf = 9.02 inch; thickness, tf = 0.77 inch
Bolts: M20; A490M do = 0.87 inch
No. of rows = 2; Nos of Bolt/row = 5
$e_0 = 1.26$ inch; $\Omega = 1.67$ ASD (yield); $\Omega = 2$ ASD (rupture)
Gross area, Ag = 6.95 inch²; Net area, An = 5.61 inch²
$F_{nBM} = 36$ ksi; $F_{uBM} = 65$ ksi; U = 1
Block shear, V:
2 L = 54.96 inch; 2b = 2.52 inch; Agv = 42.32 inch²
Anv = 38.97 inch²; Ant = 1.94 inch²

$$\text{Rn} = 0.6\,\text{Fu Anv} + \text{U.Fu Ant} = 1646\,\text{kips}$$

Or

$0.6\,\text{Fy Agv} + \text{U.Fu Ant} = 1040\,\text{kips}$ $V = \text{Rn}/\Omega = 1040/2 = 520\,\text{kips}$

Now we get,
1. Tensile yielding, $\text{Pn}_1 = \text{FyAg}/\Omega = 125$ kips
2. Tensile rupture, $\text{Pn}_2 = \text{FuAe}/\Omega = \text{Fu An. }U/\Omega = 182$ kips
3. Block shear = 520 kips
Full strength, F = 125 kips.

High-Strength Bolts in Slip-Critical Connections (ASD)
Slip-critical connections shall be designed to prevent slip and for the limit states of bearing-type connections.
The Slip resistance for the limit state of slip = Rn.
Bolt: **M20; A490M**
Allowable Shear, Psc = Rn/Ω; **Rn = μ. Du. h_f.Tb. n_s** (AISC manual – J3-4)

Where,
   Ω = 1.5: (ASD)
   μ = 0.3: Class a (unpainted clean mill scale)
   Du = 1.13: a multiplier factor
   Tb = 179 kN: bolt pretension
   h_f = 1: factor for filler (1 for no filler, 0.85 for two or more filler)
   n_s = **2**: number of slip planes
   lc = 21 mm: $(e_0 - d_0)$ [clear distance between edge of hole and the edge of adjacent hole or edge of the material].

$$Rn = 0.3 \times 1.13 \times 1 \times 179 \times 2 = 121.36 \, kN$$

$$\textbf{Psc} = Rn \, / \, \Omega = 121.36 \, / \, 1.5 = \textbf{81 kN}$$

Slip resistance for bearing, Pb = Rn/Ω
   d = 20 mm; t = **19.56** mm; Ω = 2

$$1.2 \, lc.t.Fu = 1.2 \times 21 \times 19.56 \times 450 \, / \, 1000 = 222 \, kN$$

$$2.4 \, d.t.Fu = 2.4 \times 20 \times 19.56 \times 450 \, / \, 1000 = 422 \, kN$$

Rn = 222 kN; **Pb** = Rn/Ω = **111 kN**
   Design strength = 81 kN = 18 kips (lower of Psc and Pb)
   No. of bolts per flange / side = 2 × 5 = 10
   Bolt value = 18 kips; Shear Resistantnce = 10 × 18 = **180** kips > **F = 125 kips; Safe**.
f) **Web Cover Plate and Bolt**
Member section: W24 × 84; Bolt:M20; A490 M
   Member web – depth, d = 24.1 inch; thickness, tw = 0.47 inch
   Cover plate – depth, d_{wc} = 19 inch; thickness, t_{wc} = 0.38 inch
   Width, Bw = 10 inch; Nos = 2
   Cross sectional area of web, Aw = 24 × 0.47 = 11.33 inch^2
   Required sectional area of cover plate, A required = Aw × 1.05 = 11.9 1inch^2
   Total sectional area of cover plate, A provided = 2 × 19 × 0.38 = 14.44 inch^2 > A required; Okay.
   Lateral shear, V = 0.6 FyAw/Ω; Ω = 1.67
   V = 0.5 × 0.6 × 36 × 24.1 × 0.47/1.67 = 73 kips [331 kN]

Try with bolt group **10G** (Refer Figure 4.24: **Group 10G** with 10 bolts per side) and

Bolt size **M20; A490M**; Bolt dia = **20** mm

Force / bolt, F1:

Nos / side, n_0 = 10; m = 60 mm; n = 105 mm

$$I = 142243 \quad \text{mm}^2$$

$$Z = 165 \, \text{mm}$$

$$V = 331 \, \text{kN}$$

$$e_1 = 32 \, \text{mm}$$

$$e = 62 \, \text{mm} \, (0.5 \times 60 + 32)$$

$$\cos \phi = 0.218$$

$$F_1 = 331 \times 62 \times 165 / 142243 = 24 \, \text{kN} \quad F_2 = 331 / 10 = 33 \, \text{kN}$$

$$\mathbf{F} = \sqrt{(24^2 + 33^2 + 2 \times 24 \times 33 \times 0.218)} = 45 \, \text{kN} = \mathbf{10 \, kips}$$

High-Strength Bolts in Slip-Critical Connections (ASD)

Slip-critical connections shall be designed to prevent slip and for the limit states of bearing-type connections.

Bolt: **M20; A490M**

Allowable Shear, $\mathbf{Psc = Rn / \Omega}$ $\mathbf{Rn = \mu.Du.h_f.Tb.n_s}$ (AISC manual – J3-4)

where,

Ω = 1.5 (ASD)

μ = 0.3 Class a (unpainted clean mill scale)

Du = 1.13 a multiplier factor

Tb = 179 kN bolt pretension

h_f = 1 factor for filler (1 for no filler, 0.85 for two or more filler)

n_s = **2** number of slip planes

lc = 21 mm $(e_0 - d_0)$ clear distance between edge of hole and the edge of adjacent hole or edge of the material

$$Rn = 0.3 \times 1.13 \times 1 \times 179 \times 2 = 121.36 \, \text{kN}$$

$$\mathbf{Psc} = Rn / \Omega = 121.36 / 1.5 = \mathbf{81 \, kN}$$

Slip resistance for bearing, Pb = Rn/Ω

d = 20 mm; t = **11.94** mm; Ω = 2

$$1.2 \, \text{lc.t.Fu} = 1.2 \times 21 \times 11.94 \times 450 / 1000 = 135 \, \text{kN}$$

$$2.4 \, \text{d.t.Fu} = 2.4 \times 20 \times 11.94 \times 450 / 1000 = 258 \, \text{kN}$$

$$Rn = 135 \, kN \qquad \textbf{Pb} = Rn / \Omega = \textbf{68 kN}$$

Design strength = 68 kN = 15 kips (lower of Psc and Pb)
Maximum force per bolt, F = **10** kips
Bolt value or design strength = **15** kips **> F; Safe**.

4.6.5   BEAM AND GIRDER WITH SLIP CRITICAL BOLT – LRFD

a) **Sketch**
   Refer Figure 4.20 in example 4.6 above.
b) **Material**
   Refer Table 4.7 above.
c) **Member Description**
   Spliced member: **W24 × 84; Type – 1**
   (Refer example 4.6.4 above.)
d) **Design parameters**:
   1. Flange force, F: Full tension capacity
   2. Web shear, V: 50% of web shear capacity
   3. Sum area of cover plates is 5% more than the area of section spliced.
   4. All bolts are High tensile bolt Slip-critical connection.
e) **Flange cover plate and bolt**:
   Use same as in example 4.6.4 above.
      Flange area, Af = 6.95 inch^2

$$A \text{ required} = Af \times 1.05 = 7.3 \, inch^2$$

A provided = 8.46 inch^2 > A required.

Flange force:
   wide, bf = 9.02 inch; thickness, tf = 0.77 inch
   Bolts: M20; A490M; do = 0.87 inch
   No. of rows = 2; Nos of Bolt / row = 5
   e_0 = 1.26 inch; ϕ = 0.9 LRFD (yield); ϕ = 0.75 LRFD (rupture)
   Gross area, Ag = 6.95 inch^2; Net area, An = 5.61 inch^2
   F_{nBM} = 36 ksi; F_{uBM} = 65 ksi; U = 1
   Block shear, V:
   2 L = 54.96 inch; 2b = 2.52 inch; Agv = 42.32 inch^2
   Anv = 38.97 inch^2; Ant = 1.94 inch^2

$$Rn = 0.6 \, Fu \, Anv + U.Fu \, Ant = 1646 \, kips \quad \phi = 0.75$$

Or

$$0.6 \, Fy \, Agv + U.Fu.Ant = 1040 \, kips \quad V = \phi \, Rn = 0.75 \times 1040 = 780 \, kips$$

Now we get,

1. Tensile yielding, $Pn_1 = \phi\, FyAg = 225$ kips
2. Tensile rupture, $Pn_2 = \phi\, FuAe = \phi\, Fu\, An$. $U = 273$ kips
3. Block shear $= 780$ kips

Full strength, F = 225 kips. (lower of values in 1, 2, and 3 above)

High-Strength Bolts in Slip-Critical Connections (LRFD)

Slip-critical connections shall be designed to prevent slip and for the limit states of bearing-type connections.

The Slip resistance for the limit state of slip = Rn.

Bolt: **M20 A490M**

Allowable Shear, **Psc $= \phi\, Rn; Rn = \mu.Du.h_f\,Tb.n_s$** (AISC manual – J3-4)
where,

$\phi = 1.0$: (LRFD)
$\mu = 0.3$: Class a (unpainted clean mill scale)
$Du = 1.13$: a multiplier factor
$Tb = 179$ kN: bolt pretension
$h_f = 1$: factor for filler (1 for no filler, 0.85 for two or more filler)
$n_s = 2$: number of slip planes
$lc = 21$ mm: $(e_0 - d_0)$ [clear distance between edge of hole and the edge of adjacent hole or edge of the material].

$$Rn = 0.3 \times 1.13 \times 1 \times 179 \times 2 = 121\,kN$$

$$\textbf{Psc} = \phi\, Rn = 1 \times 121 = \textbf{121\ \ kN}$$

Slip resistance for bearing, Pb $= \phi$. Rn
$d = 20$ mm; t $= \textbf{19.56}$ mm; $\phi = 0.75$

$$1.2\,lc.t.Fu = 1.2 \times 21 \times 19.56 \times 450 / 1000 = 222\,kN$$

$$2.4\,d.t.Fu = 2.4 \times 20 \times 19.56 \times 450 / 1000 = 422\,kN$$

$Rn = 222$ kN; **Pb $= \phi$. Rn = 167 kN**
Design strength = 121 kN = 27 kips (lower of Psc and Pb)
No. of bolts per flange / side $= 2 \times 5 = 10$
Bolt value or Design strength $= 27$ kips

$$\text{Shear Resistantnce} = 10 \times 27 = \textbf{270}\ \text{kips} > \textbf{F} = \textbf{225 kips;Safe.}$$

f) Web Cover Plate and Bolt

Member section: W24 × 84; Bolt:**M20; A490 M**

Member web – depth, $d = 24.1$ inch; thickness, tw $= 0.47$ inch
Cover plate – depth, $d_{wc} = 19$ inch, thickness, $t_{wc} = 0.38$ inch
Width, Bw $= 10$ inch; Nos $= 2$
Cross sectional area of web, Aw $= 24 \times 0.47 = 11.33$ inch^2

Required sectional area of cover plate, A required $= Aw \times 1.05 = 11.9$ 1nch^2

Total sectional area of cover plate, A provided $= 2 \times 19 \times 0.38 = 14.44$ inch^2 > A required. Okay.

Lateral shear, $V = \phi.\ 0.6\ FyAw;\ \phi = 0.9$

$V = 0.9 \times 0.5 \times 0.6 \times 36 \times 24.1 \times 0.47 = \mathbf{110}$ kips [499 kN]

Try with bolt group **10G** (Refer Figure 4.24: Group 10G with 10 bolts per side) and

Bolt size **M20**; **A490M**; Bolt dia. $= \mathbf{20}$ mm

Force/bolt, F1:

Nos/side, $n_0 = 10$; m $= 60$ mm; n $= 105$ mm

$I = 142243$ mm^2

$Z = 165$ mm

$V = \mathbf{499}$ kN

$e_1 = 32$ mm

$e = 62$ mm $(0.5 \times 60 + 32)$

$\cos \phi = 0.218$

$$F_1 = 499 \times 62 \times 165 / 142243 = 36 \text{ kN} \quad F_2 = 499 / 10 = 50 \text{ kN}$$

$$\mathbf{F} = \sqrt{\left(36^2 + 50^2 + 2 \times 36 \times 50 \times 0.218\right)} = 68 \text{ kN} = \mathbf{15 \text{ kips}}$$

High-Strength Bolts in Slip-Critical Connections (LRFD)

Slip-critical connections shall be designed to prevent slip and for the limit states of bearing-type connections.

Bolt: **M20; A490M**

Allowable Shear, Psc $= \phi\ Rn$; $\mathbf{Rn} = \mathbf{\mu.Du.h_f.Tb.n_s}$ (AISC manual – J3-4) where,

$\phi = 1$: (LRFD)

$\mu = 0.3$: Class a (unpainted clean mill scale)

Du $= 1.13$: a multiplier factor

Tb $= 179$ kN: bolt pretension

$h_f = 1$: factor for filler (1 for no filler, 0.85 for two or more filler)

$n_s = \mathbf{2}$: number of slip planes

lc $= 21$ mm $\quad (e_0 - d_0)$ [clear distance between edge of hole and the edge of adjacent hole or edge of the material].

$$Rn = 0.3 \times 1.13 \times 1 \times 179 \times 2 = 121 \text{ kN}$$

$$\mathbf{Psc} = \phi.Rn = 1 \times 121 = \mathbf{121 \text{ kN}}$$

Slip resistance for bearing, Pb = ϕ. Rn

d $= 20$ mm; t $= \mathbf{11.94}$ mm; $\phi = 0.75$

$$1.2 \text{ lc.t.Fu} = 1.2 \times 21 \times 11.94 \times 450 / 1000 = 135 \text{ kN}$$

$$2.4 \, d.t.Fu = 2.4 \times 20 \times 11.94 \times 450 / 1000 = 258 \, kN$$

Rn = 135 kN; **Pb = φ.**; Rn = **101 kN**
Design strength = 101 kN = 22 kips (lower of Psc and Pb)
Maximum force per bolt, F = **15 kips**
Bolt value or design strength = **22 kips > F; Safe.**

4.6.6   Column Splice with Friction Grip Bolt – IS 800 (Working Stress)

a) **Sketch**

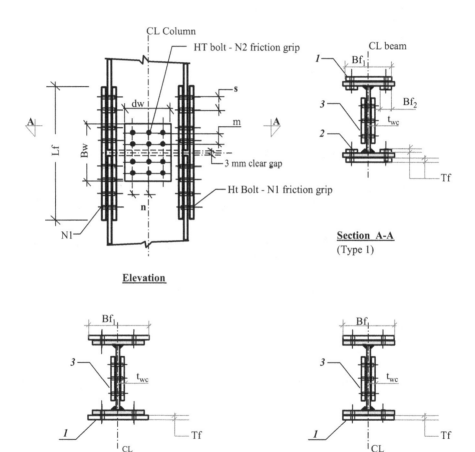

FIGURE 4.36   Column splice – bolted.

b) **Material**

Refer Table 4.6.

c) **Member Descriptions**

Spliced member: **Column C1 Type – 1**

Flange plates:	Web plate:
wide, bf = 600 mm	depth, d = 936 mm
thickness, tf = 32 mm	thickness, tw = 16 mm

Connection Bolt: **M27**; HT Bolt 8.8 grade – Friction grip

d) **Design Parameters**

1. Axial Compression, P = Pf + Pw

   a) Flange force, F: Max. of tension /comp = 0.6 fy. Af, where, Af = flange area

   b) Force on web, Pw: Full compression = 0.6 fcd. Aw, where, Aw = web area

2. Web shear, V: 2.5% of Compressive strength of column section Sum area of cover plates is 5% more than the area of section spliced.

3. Sum area of cover plates is 5% more than the area of section spliced.

4. All bolts are 8.8 grade HT bolts – Friction Grip connection.

e) **Sizing of Flange Cover Plate and Connection Bolt**:

Let us consider following dimensions for flange cover plates:

Flange cover plate -1:

wide, Bf_1 = 600 mm; thick, Tf =20 mm; length, Lf = 1100 mm

No. of plate per flange = 1

Flange cover plate 2:

wide, Bf_2 = 270 mm; thick, Tf =20 mm; length, Lf = 1100 mm

No. of plate per flange = 2

Cross sectional area of girder flange, Af = 600 × 32 = 19200 mm²

Sectional area of cover plate, A required = Af × 1.05 = 19200 × 1.05 = 20160 mm²

Total sectional area of cover plates:

A provided = 1 × 600 × 20 + 2 × 270 × 20 = 22800 mm² > A required; Okay.

Flange connection bolts, N1:

Bolt: M27; HT 8.8 grade, friction grip

Bolt dia., d = 27 mm

Hole dia., do =30 mm

Edge dist., e_0 = 51 mm

No. of rows = **4**

Nos of Bolt/row = **6**

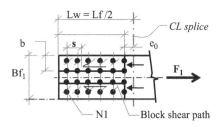

FIGURE 4.37   Flange cover plate – plan.

Lw = 1100/2 = 550 mm

Spacing of bolt, s = (550 –51 × 2) / (6 – 1) = **89.6** mm > 2.5 d; Okay.

Tension in flange plate:

The actual tensile stress on the flange plate shall be smaller of the following:

    a)   Yielding of gross section, fat = 0.6 fy (IS 800:2007 – 11.2.1)

    b)   Rupture of the net section, fat = 0.69 Tdn/Ag

    c)   Block shear resistance, fat = 0.69 Tdb/Ag

a)   by yielding:

$$\text{fat} = 0.6 \times 250 = \mathbf{150}\,\text{MPa}$$

b)   by rapture: (IS 800:2007 – 6.3.1)

$$\text{An} = 600 \times 32 - 4 \times 30 = 19080\,\text{mm}^2$$

$$\text{Tdn} = 0.9\,\text{An}\,\text{fu} / \gamma_{m1} = 0.9 \times 19080 \times 410 / 1.25 / 1000 = 5632\,\text{kN}$$

$$\text{fat} = 0.69 \times 5632000 / 19080 = \mathbf{204}\,\text{MPa}$$

c)   by block shear:

Lw = 550 mm; b = 119 mm; t = 32 mm

$$\text{Ag} = 600 \times 32 = 19200\,\text{mm}^2$$

Tb = Smaller of Tdb1 or Tdb2 (IS 800:2007 – 6.4.1; 6.4.2)

$$\boldsymbol{Tdb1 = \left[Avg.fy \,/\left(\sqrt{3}\,\gamma_{mo}\right) + 0.9\,Atn\,fu\,/\,\gamma_{m1} \right]}$$

$$\boldsymbol{Tdb2 = \left(0.9 A_{vn}\,fu \,/\left(\sqrt{3}\,\gamma_{m1}\right) + A_{tg}\,fy\,/\,\gamma_{mo} \right)}$$

Avg = gross area in shear along bolt line = $2\left(\text{Lw} - \text{e}_0\right).\text{t}$
$$= 2 \times 499 \times 32 = 31936\,\text{mm}^2$$

Avn = net area in shear along bolt line = $31936 - 2 \times 6 \times 30 \times 32 = 20416\,\text{mm}^2$

Atg = minimum gross area in tension perpendicular to the line of force
$$= 119 \times 32 = 3808\,\text{mm}^2$$

Atn = minimum net area in tension perpendicular to the line of force

$$= 3808 - 4 \times 0.5 \times 30 \times 32 = 1888\,\text{mm}^2$$

$$\text{Tdb1} = \left[31936 \times 250 / (1.732 \times 1.1) + 0.9 \times 1888 \times 410 / 1.25\right] / 1000 = 4748\,\text{kN}$$

$$\text{Tdb2} = \left[0.9 \times 20416 \times 410 / (1.732 \times 1.25) + 3808 \times 250 / 1.1\right] / 1000 = 4345\,\text{kN}$$

Tdb = 4345 kN; fat = 0.69 Tdb/Ag = **156** MPa
    fat for design = minimum of fat in a), b), and c) above = **150** MPa
    Flange force, Pf = $150 \times 600 \times 32/1000$ = **2880** kN
    Design of connection bolts (working stress) (IS 800:2007 – 11.6.1)
    Let us use,

M27	HT bolts 8.8 grade, Friction Grip
Bolt diameter	d = 20 mm
γ_{mb} = 1.25	Material Safety factor.
Fub = 830 MPa	Ultimate tensile stress of bolt.
Fu = 410 MPa	Ultimate tensile stress of plate.
$n_n = 0$	No. of shear planes with threads intercepting the shear plane.
$n_s = 2$	No. of shear planes without threads intercepting the shear plane.
Asb = 573 mm²	Nominal plain shank area of the bolt.
Anb = 459 mm²	Net shear area of the bolt at thread.
e_1 = 51 mm	End distance along bearing direction.
d_0 = 30 mm	Diameter of the hole.
p = **81** mm	Pitch distance along bearing direction (3 d).
kb = 0.567	[kb = smaller of e/3d0, (p/3d0 – 0.25), fub/fu, 1].

High-Strength Bolts in Friction Type Joint (IS 4000 1992)
Slip-critical connections shall be designed to prevent slip and for the limit
states of bearing-type connections.
    The Slip resistance for the limit state of slip = Rn. **Bolt M27**
    Allowable Shear, Psc = Rn / fos; Rn = μ Du h_fTb n_s (IS 4000 – CL 5.4)
where
    fos = 1.4: Factor of Safety
    μ = 0.35: Class a (unpainted clean mill scale) – Slip factor
    Du = 1: A multiplier factor
    Tb = 274.5 kN: Bolt pretension
    h_f = 1: Factor for filler (1 for no filler, 0.85 for two or more filler)
    n_s = 2: Number of slip planes

$$Rn = 0.35 \times 1 \times 1 \times 274.5 \times 2 = 192 \text{ kN}$$

$$\textbf{Psc} = Rn / fos = 192 / 1.4 = \textbf{37 kN}$$

Slip resistance for bearing, Vsb = 0.6 Vnpb
d = 27 mm, d0 = 30 mm, t = **32**mm, p = **81** mm, e_0 = 51 mm
kb = 0,567 [Kb = smaller of e/3d0, (p/3d0 – 0.25), fub/fu, 1]

$$Vpnb = 2.5 \text{ kb d t Fu} = 2.4 \times 0.567 \times 27 \times 32 \times 410 / 1000 = 482 \text{ kN}$$

$$\textbf{Vsb} = 0.6 \times 482 = \textbf{289 kN}$$

Bolt value or Slip resistance, Vsf = lower of Psc and Vsb = **137 kN**
Flange force, Pf = 2880 kN

$$\text{Number of bolts} (N1) = 4 \times 6 = 24 \text{ per side}$$

$$\text{Total Capacity} = 24 \times 137 = \textbf{3288 kN} \qquad > \textbf{Pf} = \textbf{2880 kN; Safe.}$$

f) Sizing of Web Cover Plate and Connection Bolts, N2

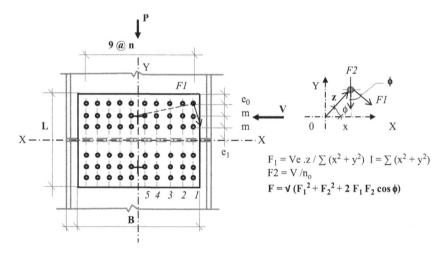

FIGURE 4.38 Group 30G.

(Refer Figures 4.21–4.29 for other groups, when applicable.)
    Column section spliced: Column **C1**
    Web cover plate: No. of plates = 2
    wide, Bw = 510 mm; depth, dw = 770 mm; thick, t_{wc} = 12 mm
    Bolt – N2
    Group = **30g**; Nos / side, n_0 = **30**; Bolt size = **M27**; Bolt dia., d = 27 mm
    e_1 = 51 mm; m = 81 mm; No. of rows / side = 3

$$n = (770 - 2 \times 51) / (30 / 3 - 1) = 74 \, \text{mm}$$

Total no. of bolts at web splice = 2 × 30 = 60
   Properties of group: Ixx = 1494682 mm²; z = 344 mm; cos ϕ = 0.236
   Web area of spliced beam, Aw = 936 × 16 = 14976 mm²

   Sectional area of cover plates = $2 \times 770 \times 12$
$$= 18480 \, \text{mm}^2 \; > \mathbf{1.05 \, Aw; \, Okay.}$$

Design forces on web connecting bolts:
   Axial compressive load, Pw = 0.6 fcd Aw (IS 800:2007 – 11.3.1)

$$Aw = 14976 \, \text{mm}^2$$

fcd = **220** MPa (∗) (IS 800:2007 – 7.1.2.1)
   [(∗) Compression due to yielding and bending about major axis (z-z);
   Unsupported length Lz = 6 m, Ly = 3m.]

$$Pw = 0.6 \times 220 \times 14976 / 1000 = 1977 \, \text{kN (vertical)}$$

No. of bolts = 30

   $\mathbf{Pw_1}$ = Force / bolt due to axial load = 1977 / 30 = **66** kN (vertical)

Lateral shear = V = 2.5% of axial compression load of column section

$$= 0.025 \times 0.6 \times 220 \times 53376 / 1000 = 176 \, \text{kN}$$

   Area of column = $2 \times 600 \times 32 + 936 \times 16 = 53376 \, \text{mm}^2$

Eccentricity, e = $e_1 + 1 \, \text{m} = 51 + 1 \times 81 = 132 \, \text{mm}$

$$F_1 = 176 \times 132 \times 344 / 1494682 = 5 \, \text{kN}$$

$$F_2 = 176 / 30 = 6 \, \text{kN}$$

$$\mathbf{F} = \sqrt{(5^2 + 6^2 + 2 \times 5 \times 6 \times 0.236)} = \mathbf{9} \, \text{kN}$$

$$\cos \phi = 0.236$$

$$\phi = 76.35^\circ$$

$$\alpha = 13.65^\circ$$

$$\cos \alpha = 0.972$$

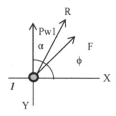

FIGURE 4.39 Force diagram.

Resultant force, $R = \sqrt{(Pw_1^2 + F^2 + 2\,Pw\,F\cos\alpha)}$

$$\mathbf{R} = \sqrt{\left(66^2 + 9^2 + 2\times 66\times 9\times 0.972\right)} = \mathbf{75}\ \text{kN}$$

High-Strength Bolts in Friction Type Joint (IS 4000 – 1992)
Slip-critical connections shall be designed to prevent slip and for the limit states of bearing-type connections.
The Slip resistance for the limit state of slip = Rn.; **Bolt M27**
Allowable Shear, Psc = Rn / fos; Rn = μ Du h_f Tb n_s (IS 4000 – CL 5.4)
Where,
   fos = 1.4 Factor of Safety
   μ = 0.35 Class a (unpainted clean mill scale) – Slip factor
   Du = 1 a multiplier factor
   Tb = 274.5 kN bolt pretension
   h_f = 1 factor for filler (1 for no filler, 0.85 for two or more filler)
   n_s = **2** number of slip planes

$$Rn = 0.35\times 1\times 1\times 274.5\times 2 = 192\ \text{kN}$$

$$\mathbf{Psc} = Rn\,/\,fos = 192\,/\,1.4 = \mathbf{137}\ \text{kN}$$

Slip resistance for bearing, Vsb = 0.6 Vnpb
   d = 27 mm, d_0 = 30 mm, t = **16** mm, p = **81** mm, e_0 = 51 mm
   kb = 0.494 [Kb = smaller of e/3d0, (p/3d0 – 0.25), fub/fu, 1]

$$Vnpb = 2.5\ kb\ d\ t\ Fu = 2.4\times 0.494\times 27\times 16\times 410\,/\,1000 = 210\ \text{kN}$$

$$\mathbf{Vsb} = 0.6\times 210 = \mathbf{126}\ \text{kN}$$

Bolt value or Slip resistance, Vsf = lower of Psc and Vsb = **126 kN** > **R = 75 kN; Safe.**

4.6.7 Column Splice with Friction Grip Bolt – IS 800 (Limit State)

a) Sketch
See Figure 4.36 in example 4.6.6.
b) Material
Refer Table 4.6
c) Member Description
Spliced member: **Column C1 Type – 1**

Flange plates:	Web plate:
wide, bf = 600 mm	depth, d = 936 mm
thickness, tf = 32 mm	thickness, tw = 16 mm

Connection Bolt: **M27** HT Bolt 8.8 grade – Friction grip
d) Design Parameters
1. Axial Compression, $P = Pf + Pw$
   a) Flange force, F: Max. of tension /comp
   b) Force on web, Pw: Full compression = 0.6 fcd. Aw, where, Aw = web area
2. Web shear, V: 2.5% of Compressive strength of column section
3. Sum area of cover plates is 5% more than the area of section spliced.
4. All bolts are 8.8 grade HT bolts friction grip connection.
e) Sizing of Flange Cover Plate and Connection Bolt:
Flange cover plate -1:
   wide, Bf_1 = 600 mm; thick, Tf =20 mm; length, Lf = 1100 mm
   No. of plate per flange = 1
Flange cover plate 2:
   wide, Bf_2 = 270 mm; thick, Tf =20 mm; length, Lf = 1100 mm
   No. of plate per flange = 2

$$A \text{ required} = 20160 \text{ mm}^2; A \text{ provided} = 22800 \text{ mm}^2$$

Flange connection bolts, N1
   Refer Figure 4.37 above (except the no. of bolts will be 7 per row)
   Bolt: M27; HT 8.8 grade, friction grip
   Bolt dia., d = 27 mm
   Hole dia., do =30 mm
   Edge dist., e_0 = 51 mm
   No. of rows = **4**
   Nos of Bolt/row = 7; **Lw = 550 mm**

Tension in flange plate:
   The actual tensile stress on the flange plate shall be smaller of the following:
   a) Yielding of gross section, fat = fy/γ_{m0} (IS 800:2007 – 6.2)
   b) Rupture of the net section, fat = Tdn/An (IS 800:2007 – 6.3)
   c) Block shear resistance, fat = Tdb/Ag

by yielding:

$$fat = 250/1.1 = \mathbf{227}\,MPa$$

by rapture: (IS 800:2007 – 6.3.1)

$$An = 600 \times 32 - 4 \times 30 = 19080\,mm^2$$

$$Tdn = 0.9\,An\,fu\,/\,\gamma_{m1} = 0.9 \times 19080 \times 410\,/\,1.25\,/\,1000 = 5632\,kN$$

$$fat = 5632000/19080 = \mathbf{295}\,MPa$$

by block shear:
Lw = 550 mm; b = 119 mm; t = 32 mm

$$Ag = 600 \times 32 = 19200\,mm^2$$

Tb = Smaller of Tdb1 or Tdb2 (IS 800:2007 – 6.4.1; 6.4.2)

$$\boldsymbol{Tdb1 = \left[Avg.fy / \left(\sqrt{3}\,\gamma_{mo} \right) + 0.9\,Atn\,fu / \gamma_{m1} \right]}$$

$$\boldsymbol{Tdb2 = \left(0.9A_{vn}fu / \left(\sqrt{3}\,\gamma_{m1} \right) + A_{tg}fy / \gamma_{mo} \right)}$$

Avg = gross area in shear along bolt line
$$= 2\left(Lw - e_0 \right).t = 2 \times 499 \times 32 = 31936\,mm^2$$

Avn = net area in shear along bolt line = $31936 - 2 \times 7 \times 30 \times 32 = 18496\,mm^2$

Atg = minimum gross area in tension perpendicular to the line of force
$$= 119 \times 32 = 3808\,mm^2$$

Atn = minimum net area in tension perpendicular to the line of force

$$= 3808 - 4 \times 0.5 \times 30 \times 32 = 1888\,mm^2$$

$$Tdb1 = \left[31936 \times 250 / \left(1.732 \times 1.1 \right) + 0.9 \times 1888 \times 410 / 1.25 \right] / 1000 = 4748\,kN$$

$$Tdb2 = \left[0.9 \times 18496 \times 410 / \left(1.732 \times 1.25 \right) + 3808 \times 250 / 1.1 \right] / 1000 = 4018\,kN$$

Tdb = 4018 kN; f_{at} = Tdb/Ag = **209** MPa
f_{at} for design = minimum of f_{at} in a), b), and c) above = **209** MPa
Flange force, Pf = 209 × 600 × 32/1000 = **4013** kN

Strength of High-Strength Bolts in Friction Type bolt for flange connection

Slip-critical connections shall be designed to prevent slip and for the limit states of bearing-type connections.

a) Slip resistance for bearing, $Vdpb = Vnpb/\gamma mf$

$d = 27$ mm; $do = 30$ mm; $t = 32$ mm; $p = 81$ mm

$e_0 = 51$ mm; $kb = 0.567$ [Kb = smaller of $e/3d0$, $(p/3d0 - 0.25)$, fub/fu, 1]

$$Vpnb = 2.5\,kb\,d\,t\,Fu = 2.4 \times 0.567 \times 27 \times 32 \times 410 \times /1000 = 482 \text{ kN}$$

$$Vdpb = 482/1.25 = 386 \text{ kN}$$

b) Slip resistance as per IS 800:2007 (IS 800:2007 – 10.4)

Bolt M27

Slip resistance, $Vsf = Vnsf/\gamma_{mf}$

where,

$\gamma_{mf} = 1.10$ at service load

$Vnsf = \mu f.\ n_e.\ Kh.\ F_0$

where,

$\mu f = 0.33$: coeff of friction

$n_e = 2$: no. of slip surface

$Kh = 1$: for fastener in clearance hole

$F_0 = 267$ kN: Pretension in bolt

$$\left(F_0 = Anb.0.7fub = 459 \times 0.7 \times 830/1000 = 267 \text{ kN}\right)$$

$$Vdsf = \left(0.33 \times 2 \times 1 \times 267\right)/1.1 = \mathbf{160} \text{ kN}$$

Bolt value = **160** kN (lower of Vdsf and Vdpb)

Flange force, Pf = **4013 kN**

Number of bolts (N1) = $4 \times \mathbf{7} = 28$

Total Capacity = $28 \times 160 = \mathbf{4486 \text{ kN}} > \mathbf{Pf};$ hence, **safe**.

c) **Sizing of Web Cover Plate and Connection Bolts, N2**:

For member section and arrangement of bolt, see Figure 4.38 above.

Column section spliced: **C1**

Web cover plate: No. of plates = 2

wide, Bw = 510 mm; depth, dw = 770 mm; thick, $t_{wc} = 12$ mm

Bolt – N2

Group = **30g**; Nos/side, n_0 = **30**; Bolt size = **M27**; Bolt dia., $d = 27$ mm

$e_1 = 51$ mm; m = 81 mm; Now of rows / side = 3

$$n = (770 - 2 \times 51)/(30/3 - 1) = 74 \, \text{mm}$$

Total no. of bolts at web splice = 2 × 30 = 60
Properties of group: Ixx = 1494682 mm²; z = 344 mm; cos ϕ = 0.236
Web area of spliced beam, Aw = 936 × 16 = 14976 mm²
Sectional area of cover plates = 2 × 770 × 12 = 18480 mm²
>1.05 Aw; Okay.

Design forces on web connecting bolts:
Axial compressive load, Pw = fcd Aw (IS 800:2007 – 7.1.2)

$$Aw = 14976 \, \text{mm}^2$$

fcd = **226** MPa (∗) (IS 800:2007 – 7.1.2.1)
    [(∗) Compression due to yielding and bending about major axis (z-z);
    Unsupported length Lz = 6 m, Ly = 3m.]

$$Pw = 220 \times 14976 / 1000 = 3385 \, \text{kN (vertical)}$$

No. of bolts = 30

$$\mathbf{Pw_1} = \text{Force / bolt due to axial load} = 3385 / 30 = \mathbf{113} \, \text{kN (vertical)}$$

Lateral shear = V = 2.5% of axial compression load of column section

$$= 0.025 \times 226 \times 53376 / 1000 = \mathbf{302} \, \text{kN}$$

$$\text{Area of column} = 2 \times 600 \times 32 + 936 \times 16 = 53376 \, \text{mm}^2$$

$$\text{Eccentricity}, e = e_1 + 1 \, \text{m} = 51 + 1 \times 81 = 132 \, \text{mm}$$

Refer Figure 4.39 above for force diagram.

$$F_1 = 302 \times 132 \times 344 / 1494682 = 9 \, \text{kN}$$

$$F_2 = 302 / 30 = 10 \, \text{kN}$$

$$\mathbf{F} = \sqrt{(9^2 + 10^2 + 2 \times 9 \times 10 \times 0.236)} = \mathbf{15} \, \text{kN}$$

$$\cos \phi = 0.236$$

$$\phi = 76.35^\circ$$

$$\alpha = 13.65^\circ$$

$$\cos\alpha = 0.972$$

Resultant force, $R = \sqrt{(Pw_1^2 + F^2 + 2\,Pw\,F\cos\alpha)}$

$$\mathbf{R} = \sqrt{\left(113^2 + 15^2 + 2 \times 113 \times 15 \times 0.972\right)} = \mathbf{128}\,\text{kN}$$

High-Strength Bolts in Friction Type Joint (IS 4000 1992)

Slip-critical connections shall be designed to prevent slip and for the limit states of bearing-type connections.

The Slip resistance for the limit state of slip = Rn.; **Bolt M27**

Slip resistance for bearing, Vdpb = Vnpb/γ_{mf}

d = 27 mm, d_0 = 30 mm, t = **16** mm, p = **81** mm, e_0 = 51 mm

kb = 0.494 [Kb = smaller of e/3d0, (p/3d0 – 0.25), fub/fu, 1]

$$\text{Vnpb} = 2.5\,\text{kb}\,\text{d}\,\text{t}\,\text{Fu} = 2.4 \times 0.494 \times 27 \times 16 \times 410 / 1000 = 210\,\text{kN}$$

$$\mathbf{Vnpb = 210}\,\text{kN}$$

$$\text{Vdpb} = \text{Vnpb} / \gamma_{mf} = 210 / 1.25 = \mathbf{168}\,\text{kN}$$

Slip resistance as per IS 800: 2007 (IS 800:2007 – 10.4)

Bolt M27 HT bolt 8.8 friction grip bolt

Slip resistance, Vsf = Vnsf/γ_{mf}; where, γ_{mf} = 1.10 at service load

$$\text{Vnsf} = \mu f . n_e . Kh . F_0$$

where,

μf = 0.33: coeff of friction

n_e = 2: no. of slip surface

Kh = 1: for fastener in clearance hole

F_0 = 267 kN: Pretension in bolt

$$\left(F_0 = \text{Anb}.0.7\text{fub} = 459 \times 0.7 \times 830 / 1000 = 267\,\text{kN}\right)$$

$$\mathbf{Vdsf} = \left(0.33 \times 2 \times 1 \times 267\right) / 1.1 = \mathbf{160}\,\text{kN}$$

Bolt value = **160** kN (lower of Vdsf and Vdpb)

Resultant force, R = 128 kN < 160 kN; Safe.

4.6.8   COLUMNS WITH SIP CRITICAL BOLT – AISC (ASD)

a) Sketch

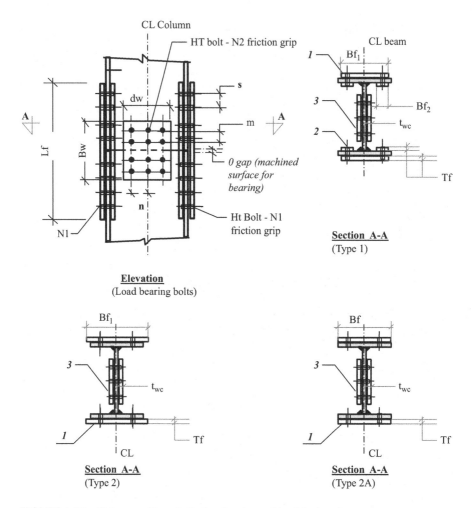

FIGURE 4.40   Column splice – bolted and ends machined for bearing.

b) Material
Refer Table 4.7.

c) Member Description
Spliced member: **Column W24 × 84 Type – 1**

Flange plates:	Web plate:
wide, bf = 9.02 inch	depth, d = 24.1 inch
thickness, tf = 0.77 inch	thickness, tw = 0.47 inch

Connection Bolt: **M20 A490M** Slip-critical connection (ASD)

d) Design Parameters

1. Axial Compression, $P = Pf + Pw$
2. The ends of column shafts are machined for bearing. It is assumed that 60% compresion will be transferred by bearing at contact face.
3. Cover plates will be designed for:
   a) Flange force, Pf: Allowable tension in Flange or 50% compression, which is more.
   b) Force on web, Pw: 40% of Full compression = 40% [Pn = Fcr. Aw], where, Aw = web area. (AISC 360-16 Eqn E3-1)
4. Web shear, $V = 30\%$ of web shear capacity (= 0.6 Fy Aw/Ω)
5. Sum area of cover plates is 5% more than the area of section spliced.
6. All bolts are High tensile bolts Slip Critical connection.

e) Flange Cover Plate and Bolt:

Flange cover plate -1:
    wide, $Bf_1 = 8$ inch; thick, Tf = 0.625 inch; length, Lf = 28 inch
    No. of plate per flange = 1
Flange cover plate 2:
    wide, $Bf_2 = 2.77$ inch; thick, Tf = 0.625 inch; length, Lf = 28 inch
    No. of plate per flange = 2
Cross sectional area of flange, Af = $9.02 \times 0.77 = 6.95$ inch^2
    Required sectional area of cover plate, A required = Af $\times 1.05 =$ **7.3** inch^2
    Total sectional area of cover plate,
    Aprovided = $1 \times 8 \times 0.625 \times 2 \times 2.77 \times 0.625 =$ **8.46** inch^2 > A required;
Okay.

f) *To Determine Flange Force and Capacity of Connection Bolts, N1*

Bolt: M20; A490M
Bolt dia., d = 20 mm = 1.00 inch
Hole dia., do =22 mm = 0.87 inch
Edge dist., e_0 = 32 mm = 1.26 inch
No. of rows = **2**
Nos of Bolt / row = **5**
Cover plate 1: Gross area, Ag = $8 \times 0.625 = 5.00$ inch^2
Net area, An = $5 - 2 \times 0.87 \times 0.625 = 3.91$ inch^2

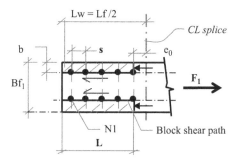

FIGURE 4.41   Flange cover plate – plan view.

Lw = 28 / 2 = 14 inch

Spacing of bolt, s = (14 – 1.26 × 2)/(5 – 1) = 2.87 inch   **s > 2.5 d; okay**.

Cover plate 1:

Tensile strength;

1. Tensile yielding, Pn1 = FyAg/Ω = 36 × 5/1.67 = **108** kips

$$\Omega = 1.67\,\text{ASD}\,\phi = 0.9\,\text{LRFD}$$

2. Tensile rupture, Pn2 = FuAe/Ω = Fu An. U/Ω = 65 × 3.91 × 1/2 = 127 kips

   Ω = 2 ASD; ϕ = 0.75 LRFD

   Ae = An U, U = shear lag factor as in Table D3.1 in AISC manual

   U = 1.00; with Case 4

Allowable Tension, T = 108 kips

To determine block shear resistance, $V = Rn/\Omega$:

L = 14 – 1.26 = 12.74 inch; 2L = 25.48 inch

t = 0.63 inch

$b = e_0$ = 1.26 inch; 2b = 2.52 inch

Rn = 0.6 Fu Anv + Ubs Fu Ant   <= 0.6 Fy Agv + Ubs Fu Ant

$$\Omega = 2\,(\text{ASD})\quad \phi = 0.75\,(\text{LRFD})$$

where,

Agv = gross area subjet to shear yielding [*L. t*] = 25.48 × 0.625 = 15.93 inch^2

Anv = net area subject to shear rupture (Agv - Σhole area) = 15.93 – 5 × 0.87 × 0.625 = 13.21 inch^2 Ant = net area subject to tension (*b. t*) = 2.52 × 0.625 = 1.58 inch^2

Ubs = 1, where tension stress is uniform (0.5, where tension stress is non-uniform)

Now, 0.6 Fu Anv + Ubs Fu Ant = $0.6 \times 65 \times 13.21 + 1 \times 65 \times 1.58 = 618$ kips

0.6 Fy Agv + Ubs Fu Ant = $0.6 \times 36 \times 15.93 + 1 \times 65 \times 1.58 = 447$ kips

Rn = 447 kips; $V = Rn/\Omega = 447/2 = 224$ kips

Block shear strength, V = 224 < Allowable tension.

Design value should be minimum of tension and block shear strength.

Maximum tension in top cover plate = **108** kips

Cover Plate 2

Pn1 = 37.32 kips; Pn2 = 39 kips; T = 37 kips

Rn = Not applicable for single row of bolts

Maximum tension in bottom cover plates = 2 × 37 = **74** kips
Total resistance of cover plates = 108 +74 = **182** kips

Flange force of spliced member W24 ×84:
Flange
wide, bf = 9.02 inch; wide, bf = 0.77 inch
Gross area, Ag = 6.95 inch^2; Net area, An = 5.61 inch^2
Bolts: M20; A490M
hole dia., do = 0.87 inch; edge dist., e_0 = 1.26 inch
no. of rows = **2**; no. of Bolt/row = **5**
Ω = 1.67 ASD (yield); Ω = 2 ASD (rapture); U = 1.0
F_{nBM} = 36 ksi; F_{uBM} = 65 ksi
Block shear, V
2L = 50.96 inch; 2b = 2.52 inch; Agv = 39.24 inch^2; Anv = 35.89 inch^2

$$Ant = 1.94 \, inch^2$$

$$Rn = 0.6 \, Fu \, Anv + U.Fu \, Ant = 1526 \, kips$$

Or,

$$0.6 \, Fy \, Agv + U.Fu \, Ant = 974 \, kips; V = Rn / \Omega = 487 \, kips$$

So, we get
1. Tensile yielding, Pn1 = FyAg/Ω = **125** kips
2. Tensile rupture, Pn2 = FuAe/Ω = Fu An. U/Ω = 182 kips
3. Block shear = 487 kips
So Flange force, **F = 125 Kips** (Full strength)

Strength of High-Strength Bolts in Slip-Critical Connections (ASD)
Bolt: M20; A490M
Slip-critical connections shall be designed to prevent slip and for the limit states of bearing-type connections.
The Slip resistance for the limit state of slip = Rn.
Allowable Shear, Psc = Rn / Ω; Rn = μ Du h$_f$Tb n$_s$ (AISC manual – J3-4)
where,

Ω = 1.5: (ASD)
μ = 0.3: Class a (unpainted clean mill scale)
Du = 1.13: a multiplier factor
Tb = 179 kN: bolt pretension
h$_f$ = 1: factor for filler (1 for no filler, 0.85 for two or more filler)
n$_s$ = **2**: number of slip planes
lc = 21 mm: (e_0 – 0.5 d_0): clear distance between edge of hole and the edge of adjacent hole or edge of the material

$$Rn = 0.3 \times 1.13 \times 1 \times 179 \times 2 = 121.36 \, kN$$

$$\textbf{Psc} = \text{Rn} / \Omega = 121.36 / 1.5 = \textbf{81} \, \text{kN}$$

Slip resistance for bearing, $\text{Pb} = \text{Rn}/\Omega$
$d = 20$ mm; $t = \textbf{19.56}$ mm; $\Omega = 2$ (AISC manual – J3-10)
Smaller of the following:

$$1.2 \, \text{lc.t.Fu} = 1.2 \times 21 \times 19.56 \times 450 / 1000 = 222 \, \text{kN}$$

$$2.4 \, \text{d. t. Fu} = 2.4 \times 20 \times 19.56 \times 450 / 1000 = 422 \, \text{kN} \qquad \text{Rn} = 222 \, \text{kN}$$

$$\text{Pb} = \text{Rn} / \Omega = 222 / 2 = 111 \, \text{kN}$$

Bolt value or Design strength = 81 kN = **18 kips** (lower of Psc and Pb)
No. of bolts per flange / side = 2 × 5 = 10

Shear Resistantnce = $10 \times 18 = \textbf{180 kips} > \text{F} = \textbf{125 kips; Safe}$.

g) Web Cover Plate and Bolt:
Member section: W24 × 84; Bolt: M20; A490 M
Member web – depth, $d = 24.1$ inch; thickness, $\text{tw} = 0.47$ inch
Cover plate – depth, $d_{wc} = 20$ inch; thickness, $t_{wc} = 0.31$ inch
Width, $\text{Bw} = 10$ inch; Nos = 2
Cross sectional area of web, $\text{Aw} = 24.1 \times 0.47 = 11.33$ inch^2
Required sectional area of cover plate, A required = $\text{Aw} \times 1.05 = 11.9$ 1inch^2
Total sectional area of cover plate, A provided = $2 \times 20 \times 0.31 = 12.4$ inch^2
> A required; Okay.
Design forces on web connecting bolts:
Axial compressive load, $\text{Pw} = 40\%$ of Pn [Pn = Fcr Aw] (AISC 360-16 Eqn E3-1)
$\text{Aw} = 11.327$ inch^2; $\Omega = 1.67$
$\text{Fcr}/\Omega = 18$ ksi (AISC 360-16 Eqn E3-1 -3-3)
[Nominal flexural strength of doubly symmetric Compact I shaped and channels member
bent about major axis, Mnx (= Mcx)]

$$\text{Pw} = 0.4 \times 18 \times 11.327 = 82 \, \text{kips}$$

Nos of bolt, $n_0 = 12$; Use Bolt Group: **Group 12G**

Pw1 = Force / bolt due to axial load = $82 / 12 = \textbf{7 kips}(\text{vertical}) = 32 \, \text{kN}$

Lateral shear, $\textbf{V} = 0.3 \times 0.6$. FyAw/$\Omega$ $\qquad\qquad [\Omega = 1.67]$

$$= 0.3 \times 0.6 \times 36 \times 24.1 \times 0.47 \times /1.67 = \textbf{44 kips} = 200 \, \text{kN}$$

Let us try with following group of bolts on each side of splice center line:

Group 12G; M20; A490M; Bolt dia = 20 mm (mkd – N2)

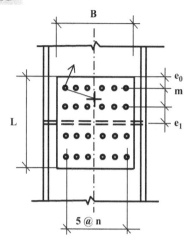

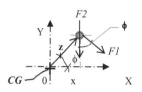

$F_1 = Ve . z / \sum (x^2 + y^2) \quad I = \sum (x^2 + y^2)$

$F2 = V / n_o$

$F = \sqrt{(F_1^2 + F_2^2 + 2\ F_1\ F_2 \cos \phi)}$

FIGURE 4.42   Group 12G.

Force/bolt, F1:

Nos/side, n_0 = 12; mkd – **N2**

I = 288035 mm^2

z = 225 mm

V = 200 kN

e_1 = 32 mm

e = 62 mm (0.5 × 60 + 32)

cos ϕ = 0.134

F_1 = 200 × 62 × 225/288035 = 10 kN

F_2 = 200/12 = 17 kN

$\mathbf{F} = \sqrt{(10^2 + 17^2 + 2 \times 10 \times 17 \times 0.134)} = \mathbf{21\ kN}$

To find out resultant force on the bolt (will be maximum on the bolt at furtherest corner of the group):

$$\cos \phi = 0.134$$

$$\phi = 82.3°$$

$$\alpha = 7.7°$$

$$\cos\alpha = 0.991$$

Resultant force, $\mathbf{R} = \sqrt{(\mathbf{Pw}_1^2 + \mathbf{F}^2 + 2\ \mathbf{Pw}\ \mathbf{F} \cos \alpha)}$

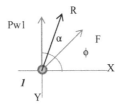

FIGURE 4.43 Force diagram.

$$\mathbf{R} = \sqrt{\left(32^2 + 21^2 + 2 \times 32 \times 21 \times 0.991\right)} = 53 \text{ kN} = \mathbf{12 \text{ kips}}$$

To determine Strength of High-Strength Bolts in Slip-Critical Connections (ASD):

Bolt: M20; A490M

Slip-critical connections shall be designed to prevent slip and for the limit states of bearing-type connections.

The Slip resistance for the limit state of slip = Rn.

Allowable Shear, Psc = Rn / Ω; Rn = μ Du h_f Tb n_s (AISC manual – J3-4)

where,

Ω = 1.5 (ASD)

μ = 0.3: Class a (unpainted clean mill scale)

Du = 1.13: a multiplier factor

Tb = 179 kN: bolt pretension

h_f = 1: factor for filler (1 for no filler, 0.85 for two or more filler)

n_s = **2**: number of slip planes

lc = 21 mm: (e_0 – 0.5 d_0) clear distance between edge of hole and the edge of adjacent hole or edge of the material

$$Rn = 0.3 \times 1.13 \times 1 \times 179 \times 2 = 121.36 \text{ kN}$$

$$\mathbf{Psc} = Rn / \Omega = 121.36 / 1.5 = \mathbf{81} \text{ kN}$$

Slip resistance for bearing, Pb = Rn/Ω

d = 20 mm; t = **12** mm; Ω = 2 (AISC manual – J3-10)

Smaller of the following:

$$1.2 \text{ lc.t.Fu} == 1.2 \times 21 \times 12 \times 450 / 1000 = 136 \text{ kN}$$

$$2.4 \text{ d.t.Fu} == 2.4 \times 20 \times 12 \times 450 / 1000 = 259 \text{ kN}; Rn = 136 \text{ kN}$$

$$Pb = Rn / \Omega = 136 / 2 = \mathbf{68} \text{ kN}$$

Bolt value or Design strength = 68 kN = **15 kips** (lower of Psc and Pb)

R = **Design strength** = **53 kN** = **12 Kips** < **Bolt value 15 kips**; **Safe**.

4.6.9   COLUMN SPLICE WITH SLIP CRITICAL BOLT – AISC (LRFD)

a) Sketch
Refer Figure 4.40.

b) Material
Refer Table 4.7.

c) Member Description
Spliced member: **Column W24 × 84 Type – 1**

Flange plates:	Web plate:
wide, bf = 9.02 inch	depth, d = 24.1 inch
thickness, tf = 0.77 inch	thickness, tw = 0.47 inch

Connection Bolt: **M20**; **A490M**; Slip-critical connection (ASD)

d) Design Parameters
Same as in example 4.6.8 above.

e) Flange Cover Plate and Bolt:
Same as in example 4.6.8 above.
  A required = 7.3 inch^2; A provided = 8.46 inch^2 > A required; Okay.

Flange force:
  wide, bf = 9.02 inch; wide, bf = 0.77 inch
  Gross area, Ag = 6.95 inch^2; Net area, An = 5.61 inch^2
  Bolts: M20; A490M
  hole dia., do = 0.87 inch; edge dist., e_0 = 1.26 inch
  no. of rows = **2**; no. of Bolt / row = **5**
  ϕ = 0.9 (yield); ϕ = 0.75 (rapture); U = 1.0
  F_{nBM} = 36 ksi; F_{uBM} = 65 ksi

Block shear, V
  2L = 50.96 inch; 2b = 2.52 inch; Agv = 39.24 inch^2; Anv = 35.89 inch^2

$$Ant = 1.94 \, inch^2$$

$$Rn = 0.6 \, Fu \, Anv + U.Fu \, Ant = 1526 \, kips$$

Or,

$$0.6 \, Fy \, Agv + U.Fu \, Ant = 974 \, kips; V = \phi.Rn = 730 \, kips$$

So, we get
1. Tensile yielding, Pn1 = ϕ. FyAg = **225** kips
2. Tensile rupture, Pn2 = ϕ. FuAe = ϕ. Fu An. U = 273 kips
3. Block shear = 730 kips

So Flange force, **F = 225 Kips** (Full strength)

Flange bolts:
Bolt: M20; A490M
Slip-critical connections shall be designed to prevent slip and for the limit states of bearing-type connections.

The Slip resistance for the limit state of slip = Rn.

Allowable Shear, Rn; Psc = ϕ.Rn = μ Du h_f Tb n_s (AISC manual – J3-4) where,

ϕ = 1.0: ($_{LRFD}$)
μ = 0.3: Class a (unpainted clean mill scale)
Du = 1.13: a multiplier factor
Tb = 179 kN: bolt pretension
h_f = 1: factor for filler (1 for no filler, 0.85 for two or more filler)
n_s = **2**: number of slip planes
lc = 21 mm: (e_0 – 0.5 d_0) clear distance between edge of hole and the edge of adjacent hole or edge of the material

$$Rn = 0.3 \times 1.13 \times 1 \times 179 \times 2 = 121\,kN$$

$$\textbf{Psc} = \phi.Rn = 1 \times 121 = \textbf{121}\,kN$$

Slip resistance for bearing, Pb = Rn/Ω
d = 20 mm; t = **19.56** mm; ϕ = 0.75; (AISC manual – J3-10)
Smaller of the following:

$$1.2\,lc.t.Fu = 1.2 \times 21 \times 19.56 \times 450 / 1000 = 222\,kN$$

$$2.4\ d.t.Fu = 2.4 \times 20 \times 19.56 \times 450 / 1000 = 422\,kN; Rn = 222\,kN$$

$$Pb = \phi.Rn = 0.75 \times 222 = \textbf{167}\,kN$$

Bolt value or Design strength = **121** kN = **27 kips** (lower of Psc and Pb)
No. of bolts per flange / side = 2 × 5 = 10

Shear Resistantnce = 10 × 27 = 270 kips > F = 225 kips; Safe.

f) Web Cover Plate and Bolt

$$Arequired = 11.9\,inch^2$$

Same as used in example 4.6.8 above.
Aprovided = 12.4 inch^2 > A required; Okay.
Design forces on web connecting bolts:
Axial compressive load, Pw = 40% of Pn [Pn = Fcr Aw] (AISC 360-16 Eqn E3-1)
Aw = 11.33 inch^2; ϕ = 0.9
ϕ. Fcr = 27 ksi (AISC 360-16 Eqn E3-1 -3-3)

[Nominal flexural strength of doubly symmetric Compact I shaped and channels member

bent about major axis, Mnx (= Mcx)]

$$Pw = 0.4 \times 27 \times 11.33 = 122 \text{ kips}$$

Nos of bolt, $n_0 = 12$; Use Bolt Group: **Group 12G**

Pw1 = Force / bolt due to axial load = $122 / 12 = \textbf{10 kips}\,(\text{vertical}) = 45\,\text{kN}$

Lateral shear, $\textbf{V} = 0.3 \times \phi \times 0.6.\ \text{FyAw}\,[\phi = 0.9]$

$$= 0.3 \times 0.9 \times 0.6 \times 36 \times 11.33 = \textbf{66 kips} = 299\,\text{kN}$$

Let us try with following group of bolts on each side of splice center line:
Group 12G; **M20**; **A490M**; **Bolt dia = 20 mm (mkd-N2)**
Refer Figure 4.42.
Force / bolt, F1:
Nos / side, $n_0 = 12$; mkd – **N2; m = 60 mm; n = 89 mm**

$$I = 288035\,\text{mm}^2$$

$$z = 225\,\text{mm}$$

$$V = 200\,\text{kN}$$

$$e_1 = 32\,\text{mm}$$

$$e = 62\,\text{mm}\,(0.5 \times 60 + 32)$$

$$\cos\phi = 0.134$$

$$F_1 = 200 \times 62 \times 225 / 288035 = 10\,\text{kN}$$

$$F_2 = 200 / 12 = 17\,\text{kN}$$

$$\textbf{F} = \sqrt{\left(10^2 + 17^2 + 2 \times 10 \times 17 \times 0.134\right)} = \textbf{21 kN}$$

To find out resultant force on the bolt (will be maximum on the bolt at furtherest corner of the group):
Refer Figure 4.43 Force diagram.

$$\cos\phi = 0.134$$

$$\phi = 82.3^\circ$$

$$\alpha = 7.7^\circ$$

$$\cos\alpha = 0.991$$

Resultant force, $\mathbf{R} = \sqrt{(\mathbf{Pw_1}^2 + \mathbf{F}^2 + 2\,\mathbf{Pw\,F}\cos\alpha)}$

$$\mathbf{R} = \sqrt{\left(45^2 + 21^2 + 2 \times 45 \times 21 \times 0.991\right)} = 66\,\text{kN} = \mathbf{15\,kips}$$

To determine Strength of High-Strength Bolts in Slip-Critical Connections (ASD):

Bolt: M20; A490M

Slip-critical connections shall be designed to prevent slip and for the limit states of bearing-type connections.

The Slip resistance for the limit state of slip = Rn.

Allowable Shear, Rn; Psc = ϕ.Rn = μ Du h_fTb n_s (AISC manual – J3-4) where,

$\phi = 1$: (LRFD)

$\mu = 0.3$: Class a (unpainted clean mill scale)

Du = 1.13: a multiplier factor

Tb = 179 kN: bolt pretension

$h_f = 1$: factor for filler (1 for no filler, 0.85 for two or more filler)

$n_s = \mathbf{2}$: number of slip planes

lc = 21 mm: $(e_0 - 0.5\,d_0)$ clear distance between edge of hole and the edge of adjacent hole or edge of the material

$$Rn = 0.3 \times 1.13 \times 1 \times 179 \times 2 = 121\,\text{kN}$$

$$\mathbf{Psc} = \phi\,Rn = 1 \times 121 = \mathbf{121}\,\text{kN}$$

Slip resistance for bearing, Pb = ϕ Rn

D = 20 mm; t = **12** mm; $\phi = 0.75$ (AISC manual – J3–10)

Smaller of the following:

$$1.2\ \text{lc.t.Fu} == 1.2 \times 21 \times 12 \times 450 / 1000 = 136\,\text{kN}$$

$$2.4\ \text{d.t.Fu} == 2.4 \times 20 \times 12 \times 450 / 1000 = 259\,\text{kN} \qquad Rn = 136\,\text{kN}$$

$$Pb = \phi\,Rn = 0.75 \times 136 = \mathbf{102}\,\text{kN}$$

Bolt value or Design strength = 102 kN = **22 kips** (lower of Psc and Pb)

R = Design strength = 66 kN = 15 Kips < Bolt value 22 kips; Safe.

REFERENCES

1. Steel Construction Manual AISC – Fourteenth edition
2. IS 800: 2007 General Construction in Steel – Code of Practice
3. AISC 360 – 10 Specification for Structural Steel Buildings

5 Design Charts

5.1 INTRODUCTION

In this section, we have given standard tables for full strength joint of rolled steel beam sections, channels, plate girders and columns, truss and bracing members with single and double angles, and tubular and rectangular hollow sections. The tables include Allowable Stress Design (ASD) and Load and Resistance Factor Design (LRFD) in accordance with AISC Steel Construction Manual and Working stress and Limit state methods as per Indian standard code for General Building and Steel Construction – IS 800. The welding types are butt weld and fillet weld. The bolted joints cover connection with bearing type bolts and slip critical bolts, as applicable. The types of details shown in these sections are commonly used in the construction of Industrial building and structure in USA, Middle Eastern countries, India, and other countries.

5.2 BEAM TO BEAM – STANDARD SHEAR CONNECTION TABLES

TABLE 5.1
Welded Joint with Web Cleat (IS 800 – 2007) – Working Stress

					Length	Fillet Weld Size		End Shear
	Beam Section			Web Cleat	Lv	ω_1	ω_2	R
SL No.	B1	B2	Nos	ISA	mm	mm	mm	kN
1	ISMB 100	ISMB 100	2	L 50 50 6	65	6	6	15
2	ISMB 125	ISMB 125	2	L 50 50 6	65	6	6	15
3	ISMB 150	ISMB 150	2	L 50 50 6	85	6	6	25
4	ISMB 175	ISMB 175	2	L 65 65 8	105	6	6	30
5	ISMB 200	ISMB 200	2	L 65 65 8	130	6	6	40
6	ISMB 250	ISMB 250	2	L 75 75 8	175	8	6	60
7	ISMB 300	ISMB 300	2	L 75 75 8	220	8	6	80
8	ISMB 350	ISMB 350	2	L 75 75 8	260	8	6	155
9	ISMB 400	ISMB 400	2	L 100 100 12	300	10	6	225
10	ISMB 450	ISMB 450	2	L 100 100 12	350	10	6	270
11	ISMB 500	ISMB 500	2	L 100 100 12	400	10	6	330
12	ISMB 600	ISMB 600	2	L 100 100 12	480	10	6	465
13	ISMC 75	ISMC 75	2	L 50 50 6	60	6	6	10
14	ISMC 100	ISMC 100	2	L 50 50 6	85	6	6	20
15	ISMC 125	ISMC 125	2	L 50 50 6	110	6	6	25
16	ISMC 150	ISMC 150	2	L 75 75 8	135	6	6	45

(Continued)

DOI: 10.1201/9781003539124-5

TABLE 5.1 (Continued)

	Beam Section		Web Cleat		Length	Fillet Weld Size		End Shear
					Lv	ω_1	ω_2	R
SL No.	B1	B2	Nos	ISA	mm	mm	mm	kN
17	ISMC 175	ISMC 175	2	L 75 75 8	160	6	6	60
18	ISMC 200	ISMC 200	2	L 75 75 8	185	6	6	75
19	ISMC 250	ISMC 250	2	L 90 90 8	235	6	6	110
20	ISMC 300	ISMC 300	2	L 90 90 8	285	6	6	140
21	ISMC 400	ISMC 400	2	L 90 90 8	385	6	6	220

Notes: Sketch: Refer Figure 3.1: Shear connection with angle cleats.
Materials: E 250 (Fe 410 W) A; fy = 250 MPa; Electrode Ex40-44XX; fu = 540 MPa, fy = 330 MPa.

TABLE 5.2
Welded Joint with Web Cleat (IS 800 – 2007) – Limit State

	Beam Section		Web Cleat		Length	Fillet Weld Size		End Shear
					Lv	ω_1	ω_2	R
SL No.	B1	B2	Nos	ISA	mm	mm	mm	kN
1	ISMB 100	ISMB 100	2	L 50 50 6	65	6	6	22
2	ISMB 125	ISMB 125	2	L 50 50 6	65	6	6	22
3	ISMB 150	ISMB 150	2	L 50 50 6	85	6	6	33
4	ISMB 175	ISMB 175	2	L 65 65 8	105	6	6	35
5	ISMB 200	ISMB 200	2	L 65 65 8	130	6	6	50
6	ISMB 250	ISMB 250	2	L 75 75 8	175	6	6	80
7	ISMB 300	ISMB 300	2	L 75 75 8	220	6	6	100
8	ISMB 350	ISMB 350	2	L 75 75 8	260	6	6	200
9	ISMB 400	ISMB 400	2	L 100 100 12	300	6	6	300
10	ISMB 450	ISMB 450	2	L 100 100 12	350	6	6	355
11	ISMB 500	ISMB 500	2	L 100 100 12	400	6	6	430
12	ISMB 600	ISMB 600	2	L 100 100 12	480	6	6	610
13	ISMC 75	ISMC 75	2	L 50 50 6	60	6	6	17
14	ISMC 100	ISMC 100	2	L 50 50 6	85	6	6	25
15	ISMC 125	ISMC 125	2	L 50 50 6	110	6	6	30
16	ISMC 150	ISMC 150	2	L 75 75 8	135	6	6	60
17	ISMC 175	ISMC 175	2	L 75 75 8	160	6	6	80
18	ISMC 200	ISMC 200	2	L 75 75 8	185	6	6	95
19	ISMC 250	ISMC 250	2	L 90 90 8	235	6	6	140
20	ISMC 300	ISMC 300	2	L 90 90 8	285	6	6	185
21	ISMC 400	ISMC 400	2	L 90 90 8	385	6	6	285

Notes: Sketch: Refer Figure 3.1: Shear connection with angle cleats.
Materials: E 250 (Fe 410 W) A; fy = 250 MPa; Electrode Ex40-44XX; fu = 540 MPa, fy = 330 MPa.

TABLE 5.3
Welded Joint with Web Cleat AISC (ASD)

SL	Beam Section			Web Cleat	Length	Fillet Weld Size		End Shear
					Lv	ω_1	ω_2	R
No.	B1	B2	Nos	Angle	inch	inch	inch	kips
1	W4 × 13	W4 × 13	2	Cut L 4 × 4(2) × 1/2	3	3/8	3/8	9
2	W5 × 16	W5 × 16	2	Cut L 4 × 4(2) × 1/2	4	1/4	1/4	9
3	W6 × 9	W6 × 9	2	L 2-1/2 × 2-1/2 × 5/16	5	1/4	1/4	8
4	W8 × 13	W8 × 13	2	L 2-1/2 × 2-1/2 × 5/16	6	1/4	1/4	15
5	W10 × 17	W10 × 17	2	L 2-1/2 × 2-1/2 × 5/16	8	1/4	1/4	20
6	W12 × 26	W12 × 26	2	L 3 × 3 × 3/8	10	1/4	1/4	23
7	W14 × 30	W14 × 30	2	L 3 × 3 × 3/8	12	1/4	1/4	30
8	W16 × 36	W16 × 36	2	L 3 × 3 × 3/8	14	1/4	1/4	35
9	W18 × 40	W18 × 40	2	L 4 × 4 × 1/2	16	1/4	1/4	45
10	W21 × 50	W21 × 50	2	L 4 × 4 × 1/2	18	1/4	1/4	65
11	W24 × 84	W24 × 84	2	L 4 × 4 × 1/2	22	3/8	3/8	95
12	C3 × 4.1	C3 × 4.1	2	Cut L 4 × 4(2) ×1/2	2	1/2	1/2	4
13	C4 × 5.4	C4 × 5.4	2	Cut L 4 × 4(2) ×1/2	3	1/2	1/2	6
14	C5 × 6.7	C5 × 6.7	2	Cut L 4 × 4(2) ×1/2	4	5/16	5/16	8
15	C6 × 8.2	C6 × 8.2	2	Cut L 4 × 4(2) ×1/2	5	5/16	5/16	10
16	C7 × 9.8	C7 × 9.8	2	L 2-1/2 × 2-1/2 × 5/16	6	1/4	1/4	12
17	C8 × 11.5	C8 × 11.5	2	L 3-1/2 × 3-1/2 × 5/16	7	1/4	1/4	15
18	C9 × 13.4	C9 × 13.4	2	L 3-1/2 × 3-1/2 × 5/16	8	1/4	1/4	18
19	C10 × 15.3	C10 × 15.3	2	L 3-1/2 × 3-1/2 × 5/16	9	1/4	1/4	20
20	C12 × 20.7	C12 × 20.7	2	L 3-1/2 × 3-1/2 × 5/16	11	1/4	1/4	25
21	C15 × 33.9	C15 × 33.9	2	L 4 × 4 × 1/2	13	3/8	3/8	50

Notes: Sketch: Refer Figure 3.1: Shear connection with angle cleats.
Materials: AISC A36; fy = 36 ksi; Electrode Fe 70; fu = 70 ksi, fy = 42 ksi.

TABLE 5.4
Welded Joint with Web Cleat AISC (LRFD)

SL	Beam Section			Web Cleat	Length	Fillet Weld Size		End Shear
					Lv	ω_1	ω_2	R
No.	B1	B2	Nos	Angle	inch	inch	inch	kips
1	W4 × 13	W4 × 13	2	Cut L 4 × 4(2) ×1/2	3	3/8	3/8	12
2	W5 × 16	W5 × 16	2	Cut L 4 × 4(2) ×1/2	4	1/4	1/4	12
3	W6 × 9	W6 × 9	2	L 2-1/2 × 2-1/2 × 5/16	5	1/4	1/4	12
4	W8 × 13	W8 × 13	2	L 2-1/2 × 2-1/2 × 5/16	6	1/4	1/4	23

(Continued)

TABLE 5.4 (Continued)

SL No.	Beam Section			Web Cleat	Length Lv	Fillet Weld Size ω_1	ω_2	End Shear R
	B1	B2	Nos	Angle	inch	inch	inch	kips
5	W10 × 17	W10 × 17	2	L 2-1/2 × 2-1/2 × 5/16	8	1/4	1/4	30
6	W12 × 26	W12 × 26	2	L 3 × 3 × 3/8	10	1/4	1/4	35
7	W14 × 30	W14 × 30	2	L 3 × 3 × 3/8	12	1/4	1/4	45
8	W16 × 36	W16 × 36	2	L 3 × 3 × 3/8	14	1/4	1/4	55
9	W18 × 40	W18 × 40	2	L 4 × 4 × 1/2	16	1/4	1/4	70
10	W21 × 50	W21 × 50	2	L 4 × 4 × 1/2	18	1/4	1/4	100
11	W24 × 84	W24 × 84	2	L 4 × 4 × 1/2	22	1/4	1/4	145
12	C3 × 4.1	C3 × 4.1	2	Cut L 4 × 4(2) ×1/2	2	3/8	3/8	6
13	C4 × 5.4	C4 × 5.4	2	Cut L 4 × 4(2) ×1/2	3	1/4	1/4	9
14	C5 × 6.7	C5 × 6.7	2	Cut L 4 × 4(2) ×1/2	4	1/4	1/4	12
15	C6 × 8.2	C6 × 8.2	2	Cut L 4 × 4(2) ×1/2	5	1/4	1/4	15
16	C7 × 9.8	C7 × 9.8	2	L 2-1/2 × 2-1/2 × 5/16	6	1/4	1/4	18
17	C8 × 11.5	C8 × 11.5	2	L 3-1/2 × 3-1/2 × 5/16	7	1/4	1/4	22
18	C9 × 13.4	C9 × 13.4	2	L 3-1/2 × 3-1/2 × 5/16	8	1/4	1/4	25
19	C10 × 15.3	C10 × 15.3	2	L 3-1/2 × 3-1/2 × 5/16	9	1/4	1/4	30
20	C12 × 20.7	C12 × 20.7	2	L 3-1/2 × 3-1/2 × 5/16	11	1/4	1/4	40
21	C15 × 33.9	C15 × 33.9	2	L 4 × 4 × 1/2	13	3/8	3/8	75

Notes: Sketch: Refer Figure 3.1: Shear connection with angle cleats.
Materials: AISC A36; fy = 36 ksi; Electrode Fe 70; fu = 70 ksi, fy = 42 ksi.

TABLE 5.5
Welded Joint with Face Plate – IS 800 (Working Stress)

SL No.	Member Section		Face Plate Length, L	Width, W	Thick, tp	Fillet Weld Weld, ω_1	Weld, ω_2	End Shear
	B1	B2	mm	mm	mm	mm	mm	kN
1	ISMB 100	ISMB 100	90	75	6	6	6	23
2	ISMB 125	ISMB 125	115	75	6	6	6	33
3	ISMB 150	ISMB 150	140	80	8	6	6	44
4	ISMB 175	ISMB 175	165	90	8	6	6	60
5	ISMB 200	ISMB 200	185	100	10	6	6	71
6	ISMB 250	ISMB 250	230	125	10	8	8	107
7	ISMB 300	ISMB 300	280	140	10	8	8	142
8	ISMB 350	ISMB 350	325	140	10	8	8	181

(*Continued*)

TABLE 5.5 (Continued)

			Face Plate			Fillet Weld		
	Member Section		Length, L	Width, W	Thick, tp	Weld, ω_1	Weld, ω_2	End Shear
SL No.	B1	B2	mm	mm	mm	mm	mm	kN
9	ISMB 400	ISMB 400	375	140	12	10	10	229
10	ISMB 450	ISMB 450	420	150	12	10	10	272
11	ISMB 500	ISMB 500	470	180	12	10	10	330
12	ISMB 600	ISMB 600	565	210	12	12	12	466
13	ISMC 75	ISMC 75	65	40	6	6	6	18
14	ISMC 100	ISMC 100	90	50	6	6	6	27
15	ISMC 125	ISMC 125	115	65	6	6	6	37
16	ISMC 150	ISMC 150	135	75	8	6	6	48
17	ISMC 175	ISMC 175	160	75	8	6	6	61
18	ISMC 200	ISMC 200	185	75	10	6	6	75
19	ISMC 250	ISMC 250	230	80	12	8	8	110
20	ISMC 300	ISMC 300	280	90	12	8	8	144
21	ISMC 400	ISMC 400	375	100	12	10	10	220

Notes: Sketch: Refer Figure 3.5: Shear connection with face plate – welded.
Materials: E 250 (Fe 410 W) A; fy = 250 MPa; Electrode Ex40-44XX; fu = 540 MPa, fy = 330 MPa.

TABLE 5.6
Welded Joint with Face Plate – IS 800 (Limit State)

			Face Plate			Fillet Weld		
	Member Section		Length, L	Width, W	Thick, tp	Weld, ω_1	Weld, ω_2	End Shear
SL No.	B1	B2	mm	mm	mm	mm	mm	kN
1	ISMB 100	ISMB 100	90	75	6	6	6	30
2	ISMB 125	ISMB 125	115	75	6	6	6	40
3	ISMB 150	ISMB 150	140	80	8	6	6	55
4	ISMB 175	ISMB 175	165	90	8	6	6	75
5	ISMB 200	ISMB 200	185	100	10	6	6	90
6	ISMB 250	ISMB 250	230	125	10	6	6	140
7	ISMB 300	ISMB 300	280	140	10	6	6	185
8	ISMB 350	ISMB 350	325	140	10	6	6	235
9	ISMB 400	ISMB 400	375	140	12	6	6	300
10	ISMB 450	ISMB 450	420	150	12	6	6	355
11	ISMB 500	ISMB 500	470	180	12	6	6	430
12	ISMB 600	ISMB 600	565	210	12	6	6	610
13	ISMC 75	ISMC 75	65	40	6	6	6	20

(Continued)

TABLE 5.6 (Continued)

			Face Plate			Fillet Weld		
			Length, L	Width, W	Thick, tp	Weld, ω_1	Weld, ω_2	End Shear
	Member Section							
SL No.	B1	B2	mm	mm	mm	mm	mm	kN
14	ISMC 100	ISMC 100	90	50	6	6	6	35
15	ISMC 125	ISMC 125	115	65	6	6	6	45
16	ISMC 150	ISMC 150	135	75	8	6	6	60
17	ISMC 175	ISMC 175	160	75	8	6	6	80
18	ISMC 200	ISMC 200	185	75	10	6	6	95
19	ISMC 250	ISMC 250	230	80	12	6	6	140
20	ISMC 300	ISMC 300	280	90	12	6	6	185
21	ISMC 400	ISMC 400	375	100	12	6	6	285

Notes: Sketch: Refer Figure 3.5: Shear connection with face plate – welded.
Materials: E 250 (Fe 410 W) A; fy = 250 MPa; Electrode Ex40-44XX; fu = 540 MPa, fy = 330 MPa.

TABLE 5.7
Welded Joint with Face Plate AISC (ASD)

			Face Plate			Fillet Weld		
			Length, L	Width, W	Thick, tp	Weld, ω_1	Weld, ω_2	End Shear
SL	Member Section							
No.	B1	B2	inch	inch	inch	inch	inch	kips
1	W4 × 13	W4 × 13	3	5	5/16	1/4	1/4	9
2	W5 × 16	W5 × 16	4	6	5/16	1/4	1/4	9
3	W6 × 9	W6 × 9	5	5	1/4	3/16	3/16	8
4	W8 × 13	W8 × 13	6	5	5/16	1/4	1/4	15
5	W10 × 17	W10 × 17	7	5	5/16	1/4	1/4	20
6	W12 × 26	W12 × 26	9	7-1/2	5/16	1/4	1/4	23
7	W14 × 30	W14 × 30	10	8	5/8	1/2	1/2	31
8	W16 × 36	W16 × 36	11	8	3/8	5/16	5/16	39
9	W18 × 40	W18 × 40	13	7	3/8	5/16	5/16	48
10	W21 × 50	W21 × 50	15	7-1/2	5/8	1/2	1/2	67
11	W24 × 84	W24 × 84	17	10	5/8	1/2	1/2	98
12	C3 × 4.1	C3 × 4.1	3	4	1/4	3/16	3/16	4
13	C4 × 5.4	C4 × 5.4	4	4	1/4	3/16	3/16	6
14	C5 × 6.7	C5 × 6.7	4	4	1/4	3/16	3/16	8
15	C6 × 8.2	C6 × 8.2	5	4	1/4	3/16	3/16	10
16	C7 × 9.8	C7 × 9.8	6	4	5/16	1/4	1/4	12
17	C8 × 11.5	C8 × 11.5	6	4	5/16	1/4	1/4	15
18	C9 × 13.4	C9 × 13.4	7	4	5/16	1/4	1/4	18

(Continued)

TABLE 5.7 (Continued)

SL	Member Section		Face Plate			Fillet Weld		
			Length, L	Width, W	Thick, tp	Weld, ω_1	Weld, ω_2	End Shear
No.	B1	B2	inch	inch	inch	inch	inch	kips
19	C10 × 15.3	C10 × 15.3	8	4	5/16	1/4	1/4	20
20	C12 × 20.7	C12 × 20.7	9	4	3/8	5/16	5/16	29
21	C15 × 33.9	C15 × 33.9	10	4-1/2	1/2	3/8	3/8	44

Notes: Sketch: Refer Figure 3.5: Shear connection with face plate – welded.
Materials: AISC A36; fy = 36 ksi; Electrode Fe 70; fu = 70 ksi, fy = 42 ksi.

TABLE 5.8
Welded Joint with Face Plate AISC (LRFD)

SL	Member Section		Face Plate			Fillet Weld		
			Length, L	Width, W	Thick, tp	Weld, ω_1	Weld, ω_2	End Shear
No.	B1	B2	inch	inch	inch	inch	inch	kips
1	W4 × 13	W4 × 13	3	5	5/16	1/4	1/4	12
2	W5 × 16	W5 × 16	4	6	5/16	1/4	1/4	12
3	W6 × 9	W6 × 9	5	5	1/4	3/16	3/16	12
4	W8 × 13	W8 × 13	6	5	5/16	1/4	1/4	23
5	W10 × 17	W10 × 17	7	5	5/16	1/4	1/4	30
6	W12 × 26	W12 × 26	9	7-1/2	5/16	1/4	1/4	35
7	W14 × 30	W14 × 30	10	8	5/8	1/4	1/4	45
8	W16 × 36	W16 × 36	11	8	3/8	1/4	1/4	55
9	W18 × 40	W18 × 40	13	7	3/8	1/4	1/4	70
10	W21 × 50	W21 × 50	15	7-1/2	5/8	1/4	1/4	100
11	W24 × 84	W24 × 84	17	10	5/8	1/4	1/4	145
12	C3 × 4.1	C3 × 4.1	3	4	1/4	3/16	3/16	6
13	C4 × 5.4	C4 × 5.4	4	4	1/4	3/16	3/16	9
14	C5 × 6.7	C5 × 6.7	4	4	1/4	3/16	3/16	12
15	C6 × 8.2	C6 × 8.2	5	4	1/4	3/16	3/16	15
16	C7 × 9.8	C7 × 9.8	6	4	5/16	1/4	1/4	18
17	C8 × 11.5	C8 × 11.5	6	4	5/16	1/4	1/4	22
18	C9 × 13.4	C9 × 13.4	7	4	5/16	1/4	1/4	25
19	C10 × 15.3	C10 × 15.3	8	4	5/16	1/4	1/4	30
20	C12 × 20.7	C12 × 20.7	9	4	3/8	1/4	1/4	40
21	C15 × 33.9	C15 × 33.9	10	4-1/2	1/2	1/4	1/4	75

Notes: Sketch: Refer Figure 3.5: Shear connection with face plate – welded.
Materials: AISC A36; fy = 36 ksi; Electrode Fe 70; fu = 70 ksi, fy = 42 ksi.

TABLE 5.9
Bolted Joint with Web Cleat (ISMB) – IS 800 (Working Stress)

SL No.	Member Section B1	Member Section B2	Web Cleat Nos	Web Cleat Size	Length Lv mm	Bolts Size	Bolts N1 Nos	Bolts N2 Nos	End Shear R kN
1	ISMB 100	ISMB 100	2	ISA 75 75 8	65	M16	1	2	15
2	ISMB 125	ISMB 125	2	ISA 75 75 8	65	M20	1	2	20
3	ISMB 150	ISMB 150	2	ISA 75 75 8	85	M24	1	2	30
4	ISMB 175	ISMB 175	2	ISA 75 75 8	105	M16	2	4	40
5	ISMB 200	ISMB 200	2	ISA 75 75 8	130	M20	2	4	50
6	ISMB 250	ISMB 250	2	ISA 75 75 10	175	M20	3	6	105
7	ISMB 300	ISMB 300	2	ISA 75 75 10	220	M20	4	8	140
8	ISMB 350	ISMB 350	2	ISA 75 75 10	260	M20	4	8	155
9	ISMB 400	ISMB 400	2	ISA 100 100 12	300	M24	4	8	195
10	ISMB 450	ISMB 450	2	ISA 100 100 12	350	M24	4	8	230
11	ISMB 500	ISMB 500	2	ISA 100 100 12	400	M24	5	10	280
12	ISMB 600	ISMB 600	2	ISA 100 100 12	480	M24	6	12	465

Notes: Sketch: Refer Figure 4.2: Shear connection with web cleats.
Materials: E 250 (Fe 410 W) A; fy = 250 MPa; HT bolt 8.8 grade (bearing type)
The above table is also valid for limit state design.

TABLE 5.10
Bolted Joint with Web Cleat (ISMC) – IS 800 (Working Stress)

SL No.	Member Section B1	Member Section B2	Web Cleat Nos	Web Cleat Size	Length Lv mm	Bolts Size	Bolts N1 Nos	Bolts N2 Nos	End Shear R kN
1	ISMC 75	ISMC 75	1	ISA 75 75 8	60	M16	1	1	10
2	ISMC 100	ISMC 100	1	ISA 75 75 8	85	M16	1	1	15
3	ISMC 125	ISMC 125	1	ISA 75 75 8	110	M16	2	2	35
4	ISMC 150	ISMC 150	1	ISA 75 75 8	135	M20	2	2	45
5	ISMC 175	ISMC 175	1	ISA 90 90 8	160	M20	2	2	60
6	ISMC 200	ISMC 200	1	ISA 90 90 8	185	M20	3	3	75
7	ISMC 250	ISMC 250	1	ISA 90 90 8	235	M20	3	3	110
8	ISMC 300	ISMC 300	1	ISA 90 90 8	285	M24	3	3	140
9	ISMC 400	ISMC 400	1	ISA 90 90 8	385	M24	4	4	220

Notes: Sketch: Refer Figure 4.2: Shear connection with web cleats.
Materials: E 250 (Fe 410 W) A; fy = 250 MPa; HT bolt 8.8 grade (bearing type).
The above table is also valid for limit state design.

TABLE 5.11
Bolted Joint with Web Cleat (W Section) – AISC (ASD)

SL	Member Section		Web Cleat		Length Lv		Bolts N1	N2	End Shear R
No.	B1	B2	Nos	Size	inch	Size	Nos	Nos	kips
1	W4 × 3	W4 × 13	2	L 2 × 2 × 3/16	3	M22	1	2	9
2	W5 × 16	W5 × 16	2	L 2 × 2 × 3/16	4	M24	1	2	9
3	W6 × 9	W6 × 9	2	L 2 × 2 × 3/16	5	M22	1	2	8
4	W8 × 13	W8 × 13	2	L 2-1/2 × 2-1/2 × 5/16	6	M16	2	4	15
5	W10 × 17	W10 × 17	2	L 2-1/2 × 2-1/2 × 5/16	8	M16	3	6	20
6	W12 × 26	W12 × 26	2	L 3 × 3 × 3/8	10	M16	4	8	23
7	W14 × 30	W14 × 30	2	L 3 × 3 × 3/8	12	M16	4	8	31
8	W16 × 36	W16 × 36	2	L 3 × 3 × 3/8	14	M16	5	10	39
9	W18 × 40	W18 × 40	2	L 4 × 4 × 1/2	16	M16	6	12	48
10	W21 × 50	W21 × 50	2	L 4 × 4 × 1/2	18	M20	6	12	67
11	W24 × 84	W24 × 84	2	L 4 × 4 × 1/2	22	M24	6	12	98

Notes: Sketch: Refer Figure 4.2: Shear connection with web cleats.
Materials: ASTM – A36; fy = 36 ksi; Bolt – A325X; Bearing type.
X = threads Excluded from shear plane; N = threads Not excluded from shear plane.
The above table is also valid for LRFD method of design.

TABLE 5.12
Bolted Joint with Web Cleat (Channel Section) – AISC (ASD)

SL	Member Section		Web Cleat		Length Lv		Bolts N1	N2	End Shear R
No.	B1	B2	Nos	Size	inch	Size	Nos	Nos	kips
1	C3 × 4.1	C3 × 4.1	1	L 4 × 4 × 1/2	2	M16	1	2	3
2	C4 × 5.4	C4 × 5.4	1	L 4 × 4 × 1/2	3	M20	1	2	4
3	C5 × 6.7	C5 × 6.7	1	L 4 × 4 × 1/2	4	M20	1	2	6
4	C6 × 8.2	C6 × 8.2	1	L 4 × 4 × 1/2	5	M20	1	2	7
5	C7 × 9.8	C7 × 9.8	1	L 4 × 4 × 1/2	6	M20	2	4	12
6	C8 × 11.5	C8 × 11.5	1	L 4 × 4 × 1/2	7	M20	3	6	15
7	C9 × 13.4	C9 × 13.4	1	L 4 × 4 × 1/2	8	M20	3	6	18
8	C10 × 15.3	C10 × 15.3	1	L 4 × 4 × 1/2	9	M22	3	6	20
9	C12 × 20.7	C12 × 20.7	1	L 4 × 4 × 1/2	11	M22	4	8	29
10	C15 × 33.9	C15 × 33.9	1	L 4 × 4 × 1/2	13	M24	5	10	51

Notes: Sketch: Refer Figure 4.2: Shear connection with web cleats.
Materials: ASTM – A36; fy = 36 ksi; Bolt – A325X; Bearing type.
X = threads Excluded from shear plane; N = threads Not excluded from shear plane.
The above table is also valid for LRFD method of design.

TABLE 5.13
Bolted Joint with Face Plate (ISMB) – IS 800 (Working Stress)

SL No.	Member Section B1	B2	Face Plate L mm	W mm	Th mm	ω_1 mm	Bolt Size	Rows	Total Nos	End Shear kN
1	ISMB 100	ISMB 100	95	75	8	6	M16	1	2	20
2	ISMB 125	ISMB 125	120	75	8	6	M16	1	2	30
3	ISMB 150	ISMB 150	145	80	8	6	M16	1	2	40
4	ISMB 175	ISMB 175	170	90	8	6	M16	2	4	60
5	ISMB 200	ISMB 200	190	100	8	6	M16	2	4	70
6	ISMB 250	ISMB 250	235	125	8	6	M16	3	6	105
7	ISMB 300	ISMB 300	285	140	8	6	M16	3	6	140
8	ISMB 350	ISMB 350	335	140	8	6	M16	4	8	180
9	ISMB 400	ISMB 400	380	140	10	6	M16	4	8	225
10	ISMB 450	ISMB 450	430	150	10	6	M16	5	10	270
11	ISMB 500	ISMB 500	475	180	10	6	M16	5	10	330
12	ISMB 600	ISMB 600	570	210	12	6	M16	6	12	465

Notes: Sketch: Refer Figure 4.2: Shear connection with web cleats.
Materials: E 250 (Fe 410 W) A; fy = 250 MPa; HT bolt 8.8 grade (bearing type).
The above table is also valid for limit state design.

TABLE 5.14
Bolted Joint with Face Plate (ISMC) – IS 800 (Working Stress)

SL No.	Member Section B1	B2	Face Plate L mm	W mm	Th mm	ω_1 mm	Bolt Size	Rows	Total Nos	End Shear kN
1	ISMC 75	ISMC 75	75	40	8	6	M16	1	2	15
2	ISMC 100	ISMC 100	95	50	8	6	M16	1	2	25
3	ISMC 125	ISMC 125	120	65	8	6	M16	1	2	35
4	ISMC 150	ISMC 150	145	75	8	6	M16	1	2	45
5	ISMC 175	ISMC 175	165	75	8	6	M16	2	4	60
6	ISMC 200	ISMC 200	190	75	8	6	M16	2	4	75
7	ISMC 250	ISMC 250	235	80	8	6	M16	3	6	110
8	ISMC 300	ISMC 300	285	90	8	6	M16	4	8	140
9	ISMC 400	ISMC 400	380	100	10	6	M16	4	8	220

Notes: Sketch: Refer Figure 4.2: Shear connection with web cleats.
Materials: E 250 (Fe 410 W) A; fy = 250 MPa; HT bolt 8.8 grade (bearing type).
The above table is also valid for limit state design.

TABLE 5.15
Bolted Joint with Face Plate (W Section) – AISC (ASD)

SL No.	Member Section B1	Member Section B2	Face Plate L inch	Face Plate W inch	Face Plate Th inch	Face Plate ω_1 inch	Bolt Size	Bolt Rows	Total Nos	End Shear kips
1	W4 × 13	W4 × 13	4	5	5/16	1/4	M16	1	2	9
2	W5 × 16	W5 × 16	5	6	5/16	1/4	M16	1	2	9
3	W6 × 9	W6 × 9	6	4	5/16	1/4	M16	1	2	8
4	W8 × 13	W8 × 13	8	5	5/16	1/4	M16	2	4	15
5	W10 × 17	W10 × 17	10	5	5/16	1/4	M16	3	6	20
6	W12 × 26	W12 × 26	12	7	5/16	1/4	M16	4	8	23
7	W14 × 30	W14 × 30	13	7	5/16	1/4	M16	4	8	30
8	W16 × 36	W16 × 36	15	7	5/16	1/4	M16	5	10	35
9	W18 × 40	W18×40	17	7	3/8	1/4	M16	5	10	45
10	W21 × 50	W21 × 50	20	7	3/8	1/4	M20	5	10	65
11	W24 × 84	W24 × 84	24	10	1/2	1/4	M24	5	10	95

Notes: Sketch: Refer Figure 4.4: Shear connection with face plate.
Materials: ASTM A36; fy = 36 ksi; Bolt: A325X; Bearing type; Electrode Fe 70xxx
X = threads Excluded from shear plane; N = threads Not excluded from shear plane.
The above table is also valid for LRFD method of design.

TABLE 5.16
Bolted Joint with Face Plate (Channel Section) – AISC (ASD)

SL No.	Member Section B1	Member Section B2	Face Plate L mm	Face Plate W mm	Face Plate Th mm	Face Plate ω_1 mm	Bolt Size	Bolt Rows	Total Nos	End Shear kN
1	C3 × 4.1	C3 × 4.1	3	2	5/16	1/4	M16	1	2	4
2	C4 × 5.4	C4 × 5.4	4	2	5/16	1/4	M16	1	2	6
3	C5 × 6.7	C5 × 6.7	5	2	5/16	1/4	M16	1	2	8
4	C6 × 8.2	C6 × 8.2	6	2	5/16	1/4	M16	1	2	10
5	C7 × 9.8	C7 × 9.8	7	3	5/16	1/4	M16	2	4	12
6	C8 × 11.5	C8 × 11.5	8	3	5/16	1/4	M16	2	4	15
7	C9 × 13.4	C9 × 13.4	9	3	5/16	1/4	M16	3	6	18
8	C10 × 15.3	C10 × 15.3	10	3	5/16	1/4	M16	3	6	20
9	C12 × 20.7	C12 × 20.7	12	3	5/16	1/4	M16	4	8	29
10	C15 × 33.9	C15 × 33.9	15	4	1/2	1/4	M20	4	8	51

Notes: Sketch: Refer Figure 4.4: Shear connection with face plate.
Materials: ASTM A36; fy = 36 ksi; Bolt: A325X; Bearing type; Electrode Fe 70xxx.
X = threads Excluded from shear plane; N = threads Not excluded from shear plane.
The above table is also valid for LRFD method of design.

5.3 BEAM TO COLUMN MOMENT CONNECTION

TABLE 5.17
Beam to Column – Full strength Welded Joint of Standard Sections [Part I]

	Plate Girder				Column				Design Force			Top Cover Plate			
	Flange Plate	Web Plate			Flange Plate		Web Plate		Moment	Shear	Axial	L1	B1	T1	S1
SL No.	mm	mm	mm	mm	mm	mm	mm	mm	kNm	kN	kN	mm	mm	mm	mm
1	600	32	1536	16	600	28	1144	28	5440	1230	75	1600	530	45	20
2	450	25	1250	16	450	25	1000	16	2750	1000	75	1250	400	40	16
3	450	25	1200	16	450	25	1000	16	2630	960	75	1200	400	40	16
4	450	20	1150	16	450	20	1000	16	2100	920	75	1150	400	32	16
5	400	20	1000	16	400	20	1000	16	1600	800	75	1000	350	32	16
6	400	20	1000	12	400	20	800	16	1500	600	75	1000	350	28	16
7	350	20	950	12	350	20	750	12	1285	570	75	950	300	32	16
8	300	16	900	12	300	16	750	12	900	540	75	765	250	25	12
9	300	16	850	10	300	16	750	12	800	425	75	850	250	25	12
10	250	16	750	10	250	16	600	12	595	375	75	750	200	25	12
11	210	20	600	12	210	20	500	12	495	360	75	600	180	28	12
12	190	16	500	10	200	16	400	12	295	250	75	500	160	25	12
13	165	16	450	10	175	16	300	12	235	225	75	450	135	25	12
14	140	16	400	8	150	16	300	12	170	160	75	400	110	25	12
1	*2*	*3*	*4*	*5*	*6*	*7*	*8*	*9*	*10*	*11*	*12*	*13*	*14*	*15*	*16*

Note: Refer Figure 3.6; Material: fy = 250MPa (36ksi); Electrode Fe 70 (70 ksi).

TABLE 5.18

Beam to Column – Full Strength Welded Joint of Standard Sections [Part II]

SL No.	Bottom Plate				Horz. Rib (6)		Diag. Rib (7)		Web Cleats (4)				Bracket (5)			Fillet Weld	
	L2	B2	T2	S2	Wide	Thk	Wide	Thk	Leg1	Leg2	Thk	Ht	Len	Ht	Thk	S3	S4
	mm	mm	mm	mm	mm	mm	mm	mm	mm	mm	mm	mm	mm	mm	mm	mm	mm
1	1600	710	32	16	250	16	0	0	100	100	10	1300	600	250	12	8	6
2	1250	500	32	16	215	12	100	6	100	100	10	1050	600	250	12	6	6
3	1200	500	32	16	200	12	100	8	100	100	10	1000	600	250	12	6	6
4	1150	500	25	16	200	12			100	100	10	950	600	250	12	6	6
5	1000	450	25	16	100	12			100	100	10	800	600	250	12	6	6
6	1000	450	25	16	100	12	100	8	100	100	8	800	600	250	12	6	6
7	950	400	28	16	170	12	100	12	75	75	6	750	300	250	10	6	6
8	765	350	20	12	125	12			75	75	6	700	300	250	10	6	6
9	850	350	20	12	100	8			75	75	6	700	300	250	10	6	6
10	750	300	20	12	100	8			75	75	6	600	250	150	10	6	6
11	600	230	25	12	100	8			65	65	6	450	200	125	8	6	6
12	500	210	20	12					65	65	6	400	200	100	8	6	6
13	450	185	20	12					65	65	6	300	200	100	8	6	6
14	400	160	16	12					65	65	6	250	200	100	8	6	6
17	*18*	*19*	*20*	*21*	*22*	*23*	*24*	*25*	*26*	*27*	*28*	*29*	*30*	*31*	*32*	*33*	*34*

TABLE 5.19
Beam to Column – Full Strength Bolted Joint of Standard Sections [Part I]

Item Mkd	Plate Girder						Column Section						Design Load	
	Flange Plate		Web Plate		Lx	Ly	Flange Plate		Web Plate		Lx	Ly	M	V
	mm	mm	mm	mm	m	m	mm	mm	mm	mm	m	m	kNm	kN
1	*2*	*3*	*4*	*5*	*6*	*7*	*8*	*9*	*10*	*11*	*12*	*13*	*14*	*15*
PG1	600	32	1536	16	6	2.5	600	28	1144	28	12	6	5440	1722
PG2	450	25	1250	16	12	2	450	25	1000	16	12	6	2765	1400
PG3	450	25	1200	16	12	2	450	25	1000	16	12	6	2630	1344
PG4	450	20	1150	16	12	2	450	20	1000	16	12	6	2100	1288
PG5	400	20	1000	16	12	2	400	20	1000	16	12	6	1615	1120
PG6	400	20	1000	12	12	2	400	20	800	16	12	6	1520	840
PG7	350	20	950	12	12	2	350	20	750	12	12	6	1285	798
PG8	300	16	900	12	12	2	300	16	750	12	12	6	900	756
PG9	300	16	850	10	10	1.5	300	16	750	12	12	6	800	595
PG10	250	16	750	10	10	1.5	250	16	600	12	9	4.5	595	525
PG11	210	20	600	12	6	1.5	210	20	500	12	9	4.5	495	504
PG12	190	16	500	10	6	1.5	200	16	400	12	6	3	295	350
PG13	165	16	450	10	6	1.5	175	16	300	12	6	3	235	315
PG14	140	16	400	8	6	1.5	150	16	300	12	6	3	170	224

Note: Refer Figure 4.8; Material: A36 (fy 36 ksi) / IS 2062 (fy 250 Mpa); HT bolt 8.8 grade; Bearing type.

TABLE 5.20
Beam to Column – Full Strength Bolted Joint of Standard Sections [Part II]

				Stiffeners		HT 8.8 Grade; Bearing Type				
				Item 5	Item 9				spacing	
Item Mkd	End Plate			Thk	Thk	Total Bolts	Dia[1] ϕ	Rows m	Vert, p	Horz, e
	height mm	width mm	thk mm	mm	mm	nos	mm	nos	mm	mm
	16	*17*	*18*	*19*	*20*	*21*	*22*	*23*	*24*	*25*
PG1	1985	625	32	16	12	50	27	2	79	125
PG2	1615	475	28	16	10	50	24	2	65	115
PG3	1550	475	28	16	10	50	24	2	62	115
PG4	1480	475	28	16	10	40	24	2	74	115
PG5	1290	425	25	16	10	40	24	2	65	115
PG6	1290	425	25	12	10	40	24	2	65	105
PG7	1230	375	25	12	8	40	20	2	62	95
PG8	1155	325	28	12	-	40	20	2	58	95
PG9	1095	325	28	10	-	40	20	2	55	90
PG10	970	275	28	10	-	30	20	2	65	90
PG11	790	235	25	12	8	20	24	2	79	105
PG12	655	215	20	10	8	20	20	2	66	90
PG13	595	190	20	10	8	20	20	2	60	90
PG14	530	165	20	8	8	20	16	2	53	75

5.4 TRUSS AND BRACING MEMBERS

5.4.1 WELDED JOINT

TABLE 5.21
Single Angle Member – Welded Joint – AISC (ASD)

SL No.	Member	Fillet Weld		Gusset Plate Thickness	Design Force, F (Axial)
	bl × bs × t	Size, ω	Lw		
	inch × inch × inch	inch	inch	inch	kips
1	L 4 × 4 × 1/2	1/4	12	5/16	80
2	L 2 × 2 × 3/16	3/16	6	5/16	15
3	L 2-1/2 × 2-1/2 × 3/16	3/16	6	5/16	19
4	L 2-1/2 × 2-1/2 × 5/16	3/16	6	5/16	30
5	L 3 × 3 × 3/16	3/16	7	5/16	23
6	L 3 × 3 × 5/16	1/4	7	5/16	35

(Continued)

TABLE 5.21 (Continued)

SL No.	Member bl × bs × t	Fillet Weld Size, ω	Lw	Gusset Plate Thickness	Design Force, F (Axial)
	inch × inch × inch	inch	inch	inch	kips
7	L 3 × 3 × 3/8	1/4	7	5/16	45
8	L 3-1/2 × 3-1/2 × 1/4	3/16	8	5/16	35
9	L 3-1/2 × 3-1/2 × 5/16	1/4	8	5/16	45
10	L 3-1/2 × 3-1/2 × 3/8	5/16	8	5/16	54
11	L 4 × 4 × 5/16	1/4	10	1/2	52
12	L 4 × 4 × 3/8	5/16	10	1/2	60
13	L 4 × 4 × 1/2	5/16	10	1/2	80
14	L 4 × 4 × 5/8	3/8	10	1/2	95
15	L 5 × 5 × 3/8	5/16	10	1/2	75
16	L 5 × 5 × 1/2	3/8	12	1/2	100
17	L 6 × 6 × 5/8	1/2	12	1/2	150
18	L 6 × 6 × 3/4	1/2	15	1/2	180
19	L 8 × 8 × 5/8	1/2	15	1/2	200
20	L 8 × 8 × 3/4	5/8	15	1/2	245

Note: Refer Figure 3.17; Material: ASTM A36; Fy = 36 ksi; Electrode – Fe 70.

TABLE 5.22
Double Angle Member – Welded Joint – AISC (ASD)

SL No.	Member bl × bs × t	Fillet Weld Size, ω	Lw	Gusset Plate Thickness	Design Force, F (Axial)
	inch × inch × inch	inch	inch	inch	kips
1	2 L 4 × 4 × 1/2	1/4	12	5/16	160
2	2 L 2 × 2 × 3/16	3/16	6	5/16	30
3	2 L 2-1/2 × 2-1/2 × 3/16	3/16	6	5/16	35
4	2 L 2-1/2 × 2-1/2 × 5/16	3/16	6	5/16	60
5	2 L 3 × 3 × 3/16	3/16	6	5/16	45
6	2 L 3 × 3 × 5/16	1/4	6	5/16	75
7	2 L 3 × 3 × 3/8	1/4	6	5/16	90
8	2 L 3-1/2 × 3-1/2 × 1/4	3/16	7	5/16	70
9	2 L 3-1/2 × 3-1/2 × 5/16	1/4	7	5/16	90
10	2 L 3-1/2 × 3-1/2 × 3/8	5/16	7	5/16	105
11	2 L 4 × 4 × 5/16	1/4	8	1/2	100
12	2 L 4 × 4 × 3/8	5/16	8	1/2	120
13	2 L 4 × 4 × 1/2	5/16	10	1/2	160
14	2 L 4 × 4 × 5/8	3/8	10	1/2	195
15	2 L 5 × 5 × 3/8	5/16	10	1/2	155

(Continued)

TABLE 5.22 (Continued)

SL No.	Member bl × bs × t inch × inch × inch	Fillet Weld Size, ω inch	Lw inch	Gusset Plate Thickness inch	Design Force, F (Axial) kips
16	2 L 5 × 5 × 1/2	3/8	10	1/2	205
17	2 L 6 × 6 × 5/8	1/2	12	1/2	305
18	2 L 6 × 6 × 3/4	1/2	15	1/2	365
19	2 L 8 × 8 × 5/8	1/2	15	1/2	415
20	2 L 8 × 8 × 3/4	5/8	15	1/2	495

Note: Refer Figures 3.17 and 3.18; Material: ASTM A36; Fy = 36 ksi; Electrode – Fe 70.

TABLE 5.23
Single Angle Member – Welded Joint – IS 800 (Working Stress)

SL No.	Member bl × bs × t mm × mm × mm	Fillet Weld Size, ω mm	Lw mm	Gusset Plate Thickness mm	Design Force, F (Axial) kN
1	ISA 50 50 6	5	150	8	85
2	ISA 65 65 6	5	200	8	110
3	ISA 65 65 8	6	200	8	145
4	ISA 75 75 6	5	225	8	130
5	ISA 75 75 8	6	225	12	170
6	ISA 75 75 10	6	275	12	210
7	ISA 90 90 6	6	225	12	155
8	ISA 90 90 8	6	275	12	205
9	ISA 90 90 10	8	275	12	255
10	ISA 100 100 8	6	300	12	230
11	ISA 100 100 10	8	300	12	285
12	ISA 100 100 12	8	350	12	335
13	ISA 110 110 10	8	350	12	315
14	ISA 110 110 12	8	375	12	375
15	ISA 110 110 16	10	375	12	460
16	ISA 130 130 10	8	375	12	375
17	ISA 130 130 12	10	350	12	445
18	ISA 150 150 16	12	450	12	640
19	ISA 150 150 20	16	400	12	760
20	ISA 200 200 16	12	570	12	865
21	ISA 200 200 20	16	520	12	1030

Note: Refer: **Figure 3.17**; Material: E250 (Fe 410W)A; Fy = 250MPa Electrode – Ex40-44XX.

TABLE 5.24
Double Angle Member – Welded Joint – IS 800 (Working Stress)

SL No.	Member bl × bs × t	Fillet Weld Size, ω	Lw	Gusset Plate Thickness	Design Force, F (Axial)
	mm × mm × mm	mm	mm	mm	kN
1	2 ISA 50 50 6	5	125	8	170
2	2 ISA 65 65 6	5	150	8	220
3	2 ISA 65 65 8	6	175	8	290
4	2 ISA 75 75 6	5	175	8	260
5	2 ISA 75 75 8	6	200	12	340
6	2 ISA 75 75 10	6	250	12	420
7	2 ISA 90 90 6	6	200	12	310
8	2 ISA 90 90 8	6	250	12	410
9	2 ISA 90 90 10	8	250	12	510
10	2 ISA 100 100 8	6	275	12	460
11	2 ISA 100 100 10	8	275	12	570
12	2 ISA 100 100 12	8	300	12	675
13	2 ISA 110 110 10	8	300	12	630
14	2 ISA 110 110 12	8	325	12	750
15	2 ISA 110 110 16	10	325	12	920
16	2 ISA 130 130 10	8	325	12	750
17	2 ISA 130 130 12	10	300	12	895
18	2 ISA 150 150 16	12	375	12	1280
19	2 ISA 150 150 20	16	375	12	1520
20	2 ISA 200 200 16	12	500	12	1730
21	2 ISA 200 200 20	16	450	12	2060

Note: Refer: **Figure 3.18**; Material: E250 (Fe 410W)A; F_y = 250MPa Electrode – Ex40-44XX.

TABLE 5.25
Bracing Members of Tubular Section – Welded Joint – IS 1161; 10748

SL No.	Nominal Dia. and Series	Tubular Section			Tubular Section			GussetPlate Thickness	Shop Weld to End Plate	Site Weld to Gusset	Design Force Axial
		Outside Dia.	Thick	Nominal Weight	Length	Wide	Thickness				
	NB	OD	Th	Wt/m	L	B	t	t_1	ω_1	ω_2	F
	mm	mm	mm	Kg	mm	mm	mm	mm	mm	mm	kN
1	50L	60.3	2.9	4.08	150	200	6	8	5	5	55
2	50M	60.3	3.6	5.03	150	200	6	8	5	5	65
3	50H	60.3	4.5	6.19	150	200	6	8	5	5	80
4	65L	76.1	3.2	5.71	175	215	6	8	5	5	80
5	65M	76.1	3.6	6.42	175	215	6	8	5	5	85
6	65H	76.1	4.5	7.93	175	215	6	8	5	5	110
7	80L	88.9	3.2	6.72	200	230	6	8	6	6	95
8	80M	88.9	4	8.36	200	230	6	8	6	6	115
9	80H	88.9	4.8	9.9	200	230	6	8	6	6	140
10	90L	101.6	3.6	8.7	210	240	8	12	6	6	120
11	90M	101.6	4	9.63	210	240	8	12	6	6	138
12	90H	101.6	4.8	11.5	210	240	8	12	6	6	160
13	100L	114.3	3.6	9.75	275	255	8	12	6	6	140
14	100M	114.3	4.5	12.2	275	255	8	12	6	6	175
15	100H	114.3	5.4	14.5	275	255	8	12	6	6	205
16	110L	127	4.5	13.6	300	265	8	12	6	6	195
17	110M	127	4.8	14.5	300	265	8	12	6	6	210
18	110H	127	5.4	16.2	300	265	8	12	6	6	235

(Continued)

TABLE 5.25 (Continued)

SL No.	Nominal Dia. and Series NB (mm)	Tubular Section Outside Dia. OD (mm)	Thick Th (mm)	Nominal Weight Wt/m (Kg)	Tubular Section Length L (mm)	Wide B (mm)	Thickness t (mm)	GussetPlate Thickness t_1 (mm)	Shop Weld to End Plate ω_1 (mm)	Site Weld to Gusset ω_2 (mm)	Design Force Axial F (kN)
19	125L	139.7	4.5	15	350	280	10	12	6	6	220
20	125M	139.7	4.8	15.9	350	280	10	12	6	6	230
21	125H	139.7	5.4	17.9	350	280	10	12	6	6	235
22	135L	152.4	4.5	16.4	350	290	10	12	6	6	240
23	135M	152.4	4.8	17.5	350	290	10	12	6	6	255
24	135H	152.4	5.4	19.6	350	290	10	12	6	6	285
25	150L	165.1	4.5	17.8	350	305	12	12	8	8	260
26	150M	165.1	4.8	18.9	350	305	12	12	8	8	275
27	150H	165.1	5.4	21.3	350	305	12	12	8	8	310
28	150L	165.1	4.5	17.8	350	305	12	12	8	8	260
29	150M	165.1	4.8	18.9	350	305	12	12	8	8	275
30	150H1	168.3	5.4	21.7	350	310	12	12	8	8	315
31	150H2	168.3	6.3	25.2	350	310	12	12	8	8	365
32	175L	193.7	4.8	22.4	350	335	12	12	8	8	325
33	175M	193.7	5.4	25.1	350	335	12	12	8	8	365
34	175H	193.7	5.9	27.3	350	335	12	12	8	8	400
35	200L	219.1	4.8	25.4	350	360	12	12	12	12	370
36	200M	219.1	5.6	29.5	350	360	12	12	12	12	430
37	200H	219.1	5.9	31	350	360	12	12	12	12	450
38	250L	273	5.9	38.9	350	415	16	16	12	12	565
39	300H	323.9	6.3	49.3	450	465	16	16	12	12	720

Note: Refer: **Figure 3.19**; Material: IS 10748; **Fy = 210 MPa** Electrode – Ex40-44XX.

TABLE 5.26

Bracing Members of Rectangular Hollow Section – Welded Joint – IS 1161; 10748

SL No.	Size	Rectangular Hollow Section			Tubular Section			Gusset Plate Thickness	Shop Weld to End Plate	Site Weld to Gusset	Design Force Axial
		Depth	Wide	Thick	Length	Wide	Thickness	t_1	ω_1	ω_2	F
		D	B	T	L	B	t				
		mm	mm	mm	mm	mm	mm	mm	mm	mm	kN
1	50 25 2.0	50	25	2	150	190	6	8	5	5	35
2	50 25 2.6	50	25	2.6	150	190	6	8	5	5	50
3	50 25 3.2	50	25	3.2	150	190	6	8	5	5	59
4	50 25 4.0	50	25	4	175	190	6	8	5	5	71
5	60 40 2.6	60	40	2.6	200	200	6	8	5	5	72
6	60 40 2.9	60	40	2.9	200	200	6	8	5	5	80
7	60 40 3.6	60	40	3.6	200	200	6	8	5	5	97
8	60 40 4.5	60	40	4.5	200	200	6	8	5	5	118
9	66 33 2.6	60	33	2.6	200	200	6	8	5	5	71
10	66 33 2.9	60	33	2.9	200	200	6	8	5	5	73
11	66 33 3.6	60	33	3.6	200	200	6	8	5	5	89
12	66 33 4.5	60	33	4.5	200	200	6	8	5	5	107
13	80 40 2.6	80	40	2.6	275	220	6	8	5	5	92
14	80 40 3.2	80	40	3.2	275	220	6	8	5	5	112
15	80 40 4	80	40	4	275	220	6	8	5	5	137
16	80 40 4.8	80	40	4.8	275	220	6	8	5	5	160
17	96 48 3.2	96	48	3.2	275	235	6	8	5	5	138
18	96 48 4	96	48	4	275	235	6	8	5	5	172

(Continued)

TABLE 5.26 (Continued)

SL No.	Size	Rectangular Hollow Section			Tubular Section			Gusset Plate Thickness t_1	Shop Weld to End Plate ω_1	Site Weld to Gusset ω_2	Design Force Axial F
		Depth D	Wide B	Thick T	Length L	Wide B	Thickness t				
		mm	mm	mm	mm	mm	mm	mm	mm	mm	kN
19	96 48 4.8	96	48	4.8	300	235	10	12	5	5	202
20	122 61 3.6	122	61	3.6	400	260	12	12	6	6	205
21	122 61 4.5	122	61	4.5	400	260	12	12	6	6	255
22	122 61 5.4	122	61	5.4	400	260	12	12	6	6	301
23	145 82 4.8	145	82	4.8	400	285	12	12	6	6	343
24	145 82 5.4	145	82	5.4	400	285	12	12	8	8	385
25	172 92 4.8	172	92	4.8	400	310	12	12	8	8	403
26	172 92 5.4	172	92	5.4	400	310	12	12	8	8	452
27	200 100 4	200	100	4	400	340	12	12	10	10	388
28	200 100 5	200	100	5	500	380	12	12	10	10	482
29	200 100 6	200	100	6	500	380	12	12	10	10	572
30	200 100 8	200	100	8	500	380	12	12	10	10	746
31	220 140 4	220	140	4	500	400	12	12	10	10	469
32	220 140 5	220	140	5	500	400	12	12	10	10	583
33	220140 6	220	140	6	500	400	12	12	10	10	694
34	220 140 8	220	140	8	600	400	16	16	12	12	909
35	240 120 4	240	120	4	600	420	16	16	10	10	469
36	240 120 5	240	120	5	600	420	16	16	10	10	583
37	240 120 6	240	120	6	600	420	16	16	10	10	694
38	240 120 8	240	120	8	600	420	16	16	10	10	909

39	**260 180 6**	260	180	6	600	460	16	16	10	10	**852**
40	**260 180 8**	260	180	8	650	460	16	16	12	12	**1125**
41	**260 180 10**	260	180	10	650	460	20	20	16	16	**1386**
42	**260 180 12**	260	180	12	700	460	20	20	16	16	**1639**
43	**300 150 6**	300	150	6	600	500	16	16	16	16	**873**
44	**300 150 8**	300	150	8	600	500	16	16	16	16	**1152**
45	**300 150 10**	300	150	10	600	500	16	16	16	16	**1420**
46	**300 150 12**	300	150	12	750	500	20	20	16	16	**1679**
47	**300 200 6**	300	200	6	650	500	16	16	16	16	**974**
48	**300 200 8**	300	200	8	650	500	16	16	16	16	**1287**
49	**300 200 10**	300	200	10	650	500	20	20	16	16	**1589**
50	**300 200 12**	300	200	12	800	550	20	20	16	16	**1882**

Note: Refer: Figure 3.19; Material: IS 10748; **Fy = 310 MPa** Electrode – Ex40-44XX.

5.4.2 Bolted Joint

TABLE 5.27
Single Angle Member Joint – Bolted – AISC (ASD)

		Bolt					Gusset Plate	
				Nom	Pitch	Edge	Thick,	Design
SL		Size	Nos	Dia., d	Dist., p	Dist., e	tg	Force, F
No.	Member Section	mm		mm	mm	mm	mm	kips
1	L 2 × 2 × 3/16	M16	3	16	48	29	8	71
2	L 2-1/2 × 2-1/2 × 3/16	M16	3	16	48	29	8	88
3	L 2-1/2 × 2-1/2 × 5/16	M16	4	16	48	29	8	143
4	L 3 × 3 × 3/16	M16	4	16	48	29	8	107
5	L 3 × 3 × 5/16	M20	4	20	60	32	8	174
6	L 3 × 3 × 3/8	M20	4	20	60	32	8	207
7	L 3-1/2 × 3-1/2 × 1/4	M20	4	20	60	32	8	166
8	L 3-1/2 × 3-1/2 × 5/16	M20	4	20	60	32	8	205
9	L 3-1/2 × 3-1/2 × 3/8	M20	4	20	60	32	12	244
10	L 4 × 4 × 5/16	M20	4	20	60	32	12	235
11	L 4 × 4 × 3/8	M20	5	20	60	32	12	280
12	L 4 × 4 × 1/2	M24	4	24	72	44	12	367
13	L 4 × 4 × 5/8	M24	5	24	72	44	12	451
14	L 5 × 5 × 3/8	M24	5	24	72	44	12	357
15	L 5 × 5 × 1/2	M24	5	24	72	44	12	468
16	L 6 × 6 × 5/8	M24	8	24	72	44	16	697
17	L 6 × 6 × 3/4	M27	8	27	81	51	16	827
18	L 8 × 8 × 5/8	M27	8	27	81	51	16	948
19	L 8 × 8 × 3/4	M30	8	30	90	57	16	1124

Note: Refer: **Figure 4.12**; Material: ASTM A 36; Fy = 36 ksi; Bolt: High tensile bearing type – A325X.

TABLE 5.28
Double Angle Member Joint – Bolted – AISC (ASD)

		Bolt					Gusset Plate	
				Nom	Pitch Dist.,	Edge Dist.,	Thick,	Design
SL		Size	Nos	Dia., d	p	e	tg	Force, F
No.	Member Section	mm		mm	mm	mm	mm	kips
1	2 L 2 × 2 × 3/16	M16	3	16	48	29	8	31
2	2 L 2-1/2 × 2-1/2 × 3/16	M16	3	16	48	29	8	39
3	2 L 2-1/2 × 2-1/2 × 5/16	M16	4	16	48	29	12	63

(Continued)

TABLE 5.28 (Continued)

SL No.	Member Section	Bolt Size mm	Nos	Nom Dia., d mm	Pitch Dist., p mm	Edge Dist., e mm	Gusset Plate Thick, tg mm	Design Force, F kips
4	2 L 3 × 3 × 3/16	M16	4	16	48	29	8	47
5	2 L 3 × 3 × 5/16	M20	4	20	60	32	12	77
6	2 L 3 × 3 × 3/8	M20	4	20	60	32	12	91
7	2 L 3-1/2 × 3-1/2 × 1/4	M20	4	20	60	32	12	73
8	2 L 3-1/2 × 3-1/2 × 5/16	M20	4	20	60	32	12	91
9	2 L 3-1/2 × 3-1/2 × 3/8	M20	5	20	60	32	12	108
10	2 L 4 × 4 × 5/16	M20	5	20	60	32	12	103
11	2 L 4 × 4 × 3/8	M20	6	20	60	32	12	123
12	2 L 4 × 4 × 1/2	M24	6	24	72	44	12	162
13	2 L 4 × 4 × 5/8	M24	5	24	72	44	20	199
14	2 L 5 × 5 × 3/8	M24	4	24	72	44	16	157
15	2 L 5 × 5 × 1/2	M24	5	24	72	44	20	207
16	2 L 6 × 6 × 5/8	M24	7	24	72	44	25	307
17	2 L 6 × 6 × 3/4	M24	8	24	72	44	28	365
18	2 L 8 × 8 × 5/8	M27	8	27	81	51	28	418
19	2 L 8 × 8 × 3/4	M27	10	27	81	51	28	496

Note: Refer: **Figure 4.14**; Material: ASTM A 36; Fy = 36 ksi; Bolt: High tensile bearing type – A325X.

TABLE 5.29
Single Angle Member Joint – Bolted – IS 800 (Working Stress)

SL No.	Member Section (mm × mm × mm)	Bolt Size mm	Nos	Nom Dia., d mm	Pitch Dist., p mm	Edge Dist., e mm	Gusset Plate Thick, tg mm	Design Force, F kN
1	ISA 50 50 6	M16	3	16	48	29	8	80
2	ISA 65 65 6	M16	4	16	48	29	8	110
3	ISA 65 65 8	M16	4	16	48	29	12	145
4	ISA 75 75 6	M20	4	20	60	32	12	130
5	ISA 75 75 8	M20	4	20	60	32	12	170
6	ISA 75 75 10	M20	4	20	60	32	12	210
7	ISA 90 90 6	M24	4	24	72	44	12	155
8	ISA 90 90 8	M24	4	24	72	44	12	205
9	ISA 90 90 10	M24	4	24	72	44	12	255

(Continued)

TABLE 5.29 (Continued)

SL No.	Member Section (mm × mm × mm)	Bolt Size mm	Nos	Bolt Nom Dia., d mm	Bolt Pitch Dist., p mm	Bolt Edge Dist., e mm	Gusset Plate Thick, tg mm	Design Force, F kN
10	ISA 100 100 8	M24	5	24	72	44	12	230
11	ISA 50 50 6	M16	3	16	48	29	8	80
12	ISA 100 100 10	M24	5	24	72	44	12	285
13	ISA 100 100 12	M24	5	24	72	44	12	335
14	ISA 110 110 10	M24	5	24	72	44	12	315
15	ISA 110 110 12	M24	5	24	72	44	12	375
16	ISA 110 110 16	M24	6	24	72	44	12	460
17	ISA 130 130 10	M24	6	24	72	44	12	375
18	ISA 130 130 12	M24	6	24	72	44	12	445
19	ISA 150 150 16	M24	8	24	72	44	12	640
20	ISA 150 150 20	M24	10	24	72	44	12	760
21	ISA 200 200 16	M27	10	27	81	51	12	865

Note: Refer: **Figure 4.12**; Material: E 250 (Fe 410 W) A; Fy = 250 MPa; Bolt: High tensile 8.8 grade bearing type.

TABLE 5.30(A)
Double Angle Member Joint – Bolted – IS 800 (Working Stress)

SL No.	Member Section (mm × mm × mm)	Bolt (Single Row) Size mm	Nos Row 1	Nos Row 2	Total	Nom Dia., d mm	Pitch Dist., p mm	Edge Dist., e mm	Gusset Plate Thick, tg mm	Design Force, F kN
1	2 ISA 50 50 6	M16	3	—	3	16	48	29	12	170
2	2 ISA 65 65 6	M16	4	—	4	16	48	29	12	220
3	2 ISA 65 65 8	M16	5	—	5	16	48	29	12	290
4	2 ISA 75 75 6	M16	5	—	5	16	48	29	12	260
5	2 ISA 75 75 8	M20	5	—	5	20	60	32	12	340
6	2 ISA 75 75 10	M20	6	—	6	20	60	32	12	420
7	2 ISA 90 90 6	M20	5	—	5	20	60	32	12	310
8	2 ISA 90 90 8	M20	6	—	6	20	60	32	12	410
9	2 ISA 90 90 10	M24	6	—	6	24	72	44	12	510
10	2 ISA 100 100 8	M24	6	—	6	24	72	44	12	460
11	2 ISA 100 100 10	M24	7	—	7	24	72	44	12	570

Note: Refer: **Figure 4.15**; Material: E 250 (Fe 410 W) A; Fy = 250 MPa; Bolt: High tensile 8.8 grade bearing type.

TABLE 5.30(B)
Double Angle Member Joint – Bolted – IS 800 (Working Stress)

SL No.	Member Section (mm × mm × mm)	Size mm	Bolt (Double Row – Staggered) Nos Row 1	Row 2	Total	Nom Dia., d mm	Pitch Dist., p mm	Edge Dist., e mm	Gusset Plate Thick, tg mm	Design Force, F kN
1	2 ISA 100 100 10	M24	4	3	7	24	72	44	12	570
2	2 ISA 100 100 12	M24	5	4	9	24	72	44	12	680
3	2 ISA 110 110 10	M24	5	4	9	24	72	44	12	635
4	2 ISA 110 110 12	M24	5	4	9	24	72	44	12	750
5	2 ISA 110 110 16	M24	6	5	11	24	72	44	12	925
6	2 ISA 130 130 10	M24	5	4	9	24	72	44	12	750
7	2 ISA 130 130 12	M24	6	5	11	24	72	44	12	895
8	2 ISA 150 150 16	M24	6	5	11	24	72	51	16	1285
9	2 ISA 150 150 20	M27	7	6	13	27	81	57	16	1525
10	2 ISA 200 200 16	M30	7	6	13	30	90	57	16	1735
11	2 ISA 200 200 20	M30	8	7	15	30	90	44	16	2065

Note: Refer: **Figure 4.16**; Material: E 250 (Fe 410 W) A; Fy = 250 MPa; Bolt: High tensile 8.8 grade bearing type.

TABLE 5.31
Tubular Section – IS 800 (Working Stress)

SL No.	NB and Series	Tube or Pipe Outside Diameter OD	Tube or Pipe Thickness Th	Tube or Pipe Nominal Weight Wt/m	Tube or Pipe Area of Cross Section A	End Plate Length L	End Plate Wide B	End Plate Thickness t	Gusset Plate Thickness t_g	Shop Weld to End Plate ω_1	HT Bolts 8.8 Grade Bolt Size Metric	Row 1	Row 2	Total Bolts
	NB	mm	mm	Kg	mm²	mm	mm	mm	mm	mm	Metric	Nos.	Nos.	Nos.
	mm													
1	50L	60.3	2.9	4.08	523	300	160	8	8	5	M16	2	2	4
2	50M	60.3	3.6	5.03	641	300	160	8	8	5	M16	2	2	4
3	50H	60.3	4.5	6.19	788	300	160	8	8	5	M16	2	2	4
4	65L	76.1	3.2	5.71	732	305	175	8	8	5	M16	2	2	4
5	65M	76.1	3.6	6.42	820	305	175	8	8	5	M16	2	2	4
6	65H	76.1	4.5	7.93	1010	305	175	8	8	5	M16	2	2	4
7	80L	88.9	3.2	6.72	861	355	190	10	10	6	M16	2	2	4
8	80M	88.9	4	8.36	1070	355	190	10	10	6	M16	2	2	4
9	80H	88.9	4.8	9.9	1270	355	190	10	10	6	M16	2	2	4
10	90L	101.6	3.6	8.7	1110	405	200	12	12	6	M16	3	3	6
11	90M	101.6	4	9.63	1230	405	200	12	12	6	M16	3	3	6
12	90H	101.6	4.8	11.5	1460	405	200	12	12	6	M16	3	3	6
13	100L	114.3	3.6	9.75	1250	455	215	12	12	6	M16	3	3	6
14	100M	114.3	4.5	12.2	1550	455	215	12	12	6	M16	3	3	6
15	100H	114.3	5.4	14.5	1850	455	215	12	12	6	M16	3	3	6
16	110L	127	4.5	13.6	1730	510	245	12	12	6	M16	3	3	6
17	110M	127	4.8	14.5	1840	510	245	12	12	6	M16	3	3	6
18	110H	127	5.4	16.2	2060	510	245	12	12	6	M16	3	3	6

19	125L	139.7	4.5	15	1910	560	280	12	12	6	M16	4	4	8
20	125M	139.7	4.8	15.9	2030	560	280	12	12	6	M16	4	4	8
21	125H	139.7	5.4	17.9	2060	560	280	12	12	6	M16	4	4	8
22	135L	152.4	4.5	16.4	2090	610	290	12	12	6	M16	4	4	8
23	135M	152.4	4.8	17.5	2220	610	290	12	12	6	M16	4	4	8
24	135H	152.4	5.4	19.6	2500	610	290	12	12	6	M16	4	4	8
25	150L	165.1	4.5	17.8	2270	660	305	12	12	6	M16	4	4	8
26	150M	165.1	4.8	18.9	2420	660	305	12	12	6	M16	4	4	8
27	150H	165.1	5.4	21.3	2710	660	305	12	12	6	M16	4	4	8
28	150L	165.1	4.5	17.8	2270	660	305	12	12	6	M16	4	4	8
29	150M	165.1	4.8	18.9	2420	660	305	12	12	6	M16	4	4	8
30	150H1	168.3	5.4	21.7	2760	675	310	12	12	6	M16	4	4	8
31	150H2	168.3	6.3	25.2	3200	675	350	12	12	6	M16	4	4	8
32	175L	193.7	4.8	22.4	2850	775	375	12	12	6	M16	5	5	10
33	175M	193.7	5.4	25.1	3200	775	375	12	12	6	M16	5	5	10
34	175H	193.7	5.9	27.3	3480	775	375	12	12	6	M16	5	5	10
35	200L	219.1	4.8	25.4	3230	875	400	12	12	6	M20	5	5	10
36	200M	219.1	5.6	29.5	3750	875	400	12	12	6	M20	5	5	10
37	200H	219.1	5.9	31	3950	875	400	12	12	6	M20	5	5	10
38	250L	273	5.9	38.9	4950	1090	460	16	16	6	M20	6	6	12
39	300H	323.9	6.3	49.3	6280	1295	500	16	16	6	M20	6	6	12

Note: Refer: **Figure 4.18**; Material: IS 10748; Fy = 210 MPa; Bolt: High tensile 8.8 grade bearing type.

TABLE 5.32
Rectangular Hollow Section – IS 800 (Working Stress)

SL No.	Size (mm × mm × mm)	Rectangular Hollow Section				End Plate					HT Bolts 8.8 Grade			
		Depth D	Wide B	Thickness T	Area of Cross Section A	Length L	Wide B	Thickness t	Gusset Plate Thickness tg	Shop Weld to End Plate ω_1	Bolt Size Metric	Row 1 Nos.	Row 2 Nos.	Total Bolts Nos.
		mm	mm	mm	mm²	mm	mm	mm	mm	mm				
1	50252.0	50	25	2	274	300	150	8	8	5	M16	2	2	4
2	50252.6	50	25	2.6	349	300	150	8	8	5	M16	2	2	4
3	50253.2	50	25	3.2	419	300	150	8	8	5	M16	2	2	4
4	50254.0	50	25	4	504	300	150	8	8	5	M16	2	2	4
5	60402.6	60	40	2.6	476	300	160	8	8	5	M16	2	2	4
6	60402.9	60	40	2.9	530	300	160	8	8	5	M16	2	2	4
7	60403.6	60	40	3.6	642	300	160	8	8	5	M16	2	2	4
8	60404.5	60	40	4.5	779	300	160	8	8	5	M16	2	2	4
9	66332.6	60	33	2.6	470	300	160	8	8	5	M16	2	2	4
10	66332.9	60	33	2.9	489	300	160	8	8	5	M16	2	2	4
11	66333.6	60	33	3.6	592	300	160	8	8	5	M16	2	2	4
12	66334.5	60	33	4.5	716	300	160	8	8	5	M16	2	2	4
13	80402.6	80	40	2.6	580	320	180	8	8	5	M16	2	2	4

14	80403.2	80	40	3.2	707	320	180	8	8	5	M16	2	2	4
15	80404	80	40	4	864	320	180	8	8	5	M16	2	2	4
16	80404.8	80	40	4.8	1014	320	200	8	8	5	M16	2	2	4
17	96483.2	96	48	3.2	854	385	215	8	8	5	M16	2	2	4
18	96484	96	48	4	1056	385	215	8	8	5	M16	2	2	4
19	96484.8	96	48	4.8	1244	385	235	8	8	5	M16	2	2	4
20	122613.6	122	61	3.6	1232	490	260	12	12	5	M16	3	3	6
21	122614.5	122	61	4.5	1526	490	260	12	12	5	M16	3	3	6
22	122615.4	122	61	5.4	1801	490	260	12	12	5	M16	3	3	6
23	145824.8	145	82	4.8	2028	580	285	12	12	5	M16	3	3	6
24	145825.4	145	82	5.4	2277	580	285	12	12	5	M16	3	3	6
25	172924.8	172	92	4.8	2383	690	310	12	12	5	M16	4	4	6
26	172925.4	172	92	5.4	2676	690	310	12	12	5	M16	4	4	8
27	2001004	200	100	4	2295	800	340	16	12	5	M20	5	5	10
28	2001005	200	100	5	2850	800	340	16	12	5	M20	5	5	10
29	2001006	200	100	6	3384	800	340	16	12	5	M20	5	5	10
30	2001008	200	100	8	4416	800	340	16	12	5	M20	5	5	10
31	2201404	220	140	4	2775	880	360	16	12	5	M20	5	5	10
32	2201405	220	140	5	3450	880	360	16	12	5	M20	5	5	10
33	2201406	220	140	6	4104	880	360	16	12	5	M20	5	5	10
34	2201408	220	140	8	5376	880	440	16	12	5	M20	5	5	10
35	2401204	240	120	4	2775	960	420	16	12	6	M20	5	5	10
36	2401205	240	120	5	3450	960	420	16	12	6	M20	5	5	10
37	2401206	240	120	6	4104	960	420	16	12	6	M20	5	5	10
38	2401208	240	120	8	5376	960	420	16	12	6	M20	5	5	12
39	2601806	260	180	6	5043	1170	500	16	12	6	M22	8	8	16

(Continued)

TABLE 5.32 (Continued)

SL No.	Size (mm × mm × mm)	Rectangular Hollow Section				End Plate			Gusset Plate Thickness	Shop Weld to End Plate	HT Bolts 8.8 Grade			
		Depth D	Wide B	Thickness T	Area of Cross Section A	Length L	Wide B	Thickness t	tg	ω_1	Bolt Size	Row 1	Row 2	Total Bolts
		mm	mm	mm	mm²	mm	mm	mm	mm	mm	Metric	Nos.	Nos.	Nos.
40	2601808	260	180	8	6656	1170	500	16	12	6	M22	8	8	16
41	26018010	260	180	10	8200	1170	500	20	12	6	M22	8	8	16
42	26018012	260	180	12	9696	1170	500	25	12	6	M22	8	8	16
43	3001506	300	150	6	5163	1350	500	25	16	6	M22	8	8	16
44	3001508	300	150	8	6816	1350	500	25	16	6	M22	8	8	16
45	30015010	300	150	10	8400	1350	500	25	16	6	M22	8	8	16
46	30015012	300	150	12	9936	1350	500	25	16	6	M22	8	8	16
47	3002006	300	200	6	5763	1350	500	25	16	6	M22	8	8	16
48	3002008	300	200	8	7616	1350	500	25	16	6	M22	8	8	16
49	30020010	300	200	10	9400	1350	500	25	16	6	M22	8	8	16
50	30020012	300	200	12	11136	1350	550	25	16	6	M22	8	8	16

Note: Refer: **Figure 4.18**; Material: IS 10748; Fy = 210 MPa; Bolt: High tensile 8.8 grade bearing type.

5.5 SPLICE JOINTS

TABLE 5.33
Welded Splice in Beams and Girders – AISC (ASD)

SL No.	Member Section	Nos/ Flng	Flange Cover Plate – 1				Nos/ Flng	Flange Cover Plate – 2				Web Cover Plate – 2 Nos				Type
			Wide Bf₁ inch	Thick Tf inch	Length Lf inch	Weld ω₁ inch		Wide Bf₂ inch	Thick Tf inch	Length Lf inch	Weld ω₁ inch	Wide Bw inch	Depth dw inch	Thick t_wc inch	Weld ω₂ inch	
1	W4 × 13	1	5.00	3/8	10	1/4						6	3	1/4	3/16	2
2	W5 × 16	1	6.00	3/8	12	1/4						6	4	1/4	3/16	2
3	W6 × 9	1	5.00	3/8	14	3/16						6	5	1/4	3/16	2
4	W8 × 13	1	5.00	3/8	12	1/4						6	6	1/4	3/16	2
5	W10 × 17	1	5.00	1/2	15	1/4						6	8	1/4	3/16	2
6	W12 × 26	1	5.00	5/8	18	1/4						6	10	1/4	3/16	2A
7	W14 × 30	1	6.00	5/8	18	5/16						6	11	1/4	3/16	2A
8	W16 × 36	1	6.00	5/8	20	5/16						7	13	1/4	3/16	2A
9	W18 × 40	1	5.00	3/4	22	5/16						7	14	1/4	3/16	2A
10	W21 × 50	1	5.50	3/4	24	5/16						9	17	5/16	1/4	2A
11	W24 × 84	1	8.00	5/8	24	5/16	2	2.77	5/8	24	1/4	10	19	3/8	1/4	1
12	PG1	1	23.00	7/8	45	5/8	2	10.00	7/8	45	5/8	23	45	1/2	5/16	1
13	PG2	1	16.00	3/4	36	1/2	2	7.00	3/4	36	1/2	19	37	1/2	5/16	1
14	PG3	1	16.00	3/4	36	1/2	2	7.00	3/4	36	1/2	18	35	1/2	5/16	1
15	PG4	1	16.00	3/4	36	1/2	2	7.00	3/4	36	1/2	17	34	1/2	5/16	1
16	PG5	1	14.00	3/4	36	1/2	2	6.00	3/4	36	1/2	15	30	1/2	5/16	1
17	PG6	1	14.00	3/4	36	1/2	2	6.00	3/4	36	1/2	15	30	3/8	1/4	1
18	PG7	1	12.00	3/4	30	1/2	2	5.00	3/4	30	1/2	14	28	3/8	1/4	1
19	PG8	1	10.00	3/4	30	1/2	2	4.00	3/4	30	1/2	14	27	3/8	1/4	1
20	PG9	1	10.00	3/4	25	1/2	2	4.00	3/4	25	1/2	13	25	5/16	1/4	1
21	PG10	1	8.00	3/4	21	1/2	2	3.00	3/4	21	1/2	11	22	5/16	1/4	1

Note: Refer Figure 3.21; Material: ASTM A36; Fy = 36 ksi; Electrode – Fe 70.

TABLE 5.34
Plate Girder Sections

Girder Mark	Flange		Web	
	Wide	Thick	Wide	Thick
	mm	mm	mm	mm
PG1	600	32	1536	16
PG2	450	25	1250	16
PG3	450	25	1200	16
PG4	450	20	1150	16
PG5	400	20	1000	16
PG6	400	20	1000	12
PG7	350	20	950	12
PG8	300	16	900	12
PG9	300	16	850	10
PG10	250	16	750	10

TABLE 5.35
Welded Splice in Channel Sections – AISC (ASD)

SL No.	Member Section	Flange Cover Plate – 1					Flange Cover Plate – 2					Web Cover Plate – 2 nos				
		Nos/Flng	Wide Bf_1 inch	Thick Tf inch	Length Lf inch	Weld ω_1 inch	Nos/Flng	Wide Bf_2 inch	Thick Tf inch	Length Lf inch	Weld ω_1 inch	Wide Bw inch	Depth dw inch	Thick t_{wc} inch	Weld ω_2 inch	Type
1	C3×4.1	1	2.00	5/16	5	1/4						4	2	5/16	3/16	3
2	C4×5.4	1	2.00	5/16	6	1/4						4	3	5/16	3/16	3
3	C5×6.7	1	2.00	3/8	7	1/4						5	4	5/16	3/16	3
4	C6×8.2	1	2.00	3/8	8	1/4						5	5	5/16	3/16	3
5	C7×9.8	1	2.50	3/8	9	1/4						6	6	5/16	3/16	3
6	C8×11.5	1	2.50	1/2	10	1/4						6	6	3/8	1/4	3
7	C9×13.4	1	2.50	1/2	10	1/4						6	7	3/8	1/4	3
8	C10×15.3	1	3.00	1/2	10	1/4						6	8	1/2	1/4	3
9	C12×20.7	1	3.00	5/8	12	1/4						6	10	1/2	1/4	3
10	C15×33.9	1	3.50	3/4	15	1/4						6	12	5/8	1/4	3

Note: Refer **Figure 3.26**; Material: ASTM A36; Fy = 36 ksi; Electrode – Fe 70.

TABLE 5.36
Welded Splice in Beams and Girders – IS 800 (Working Stress)

SL No.	Member Section	Flange Cover Plate – 1					Flange Cover Plate – 2					Web Cover Plate – 2 Nos				
		Nos/ Flng	Wide Bf_1 mm	Thick Tf mm	Length Lf mm	Weld ω_1 mm	Nos/ Flng	Wide Bf_2 mm	Thick Tf mm	Length Lf mm	Weld ω_1 mm	Wide Bw mm	Depth dw mm	Thick t_{wc} mm	Weld ω_2 mm	Type
1	ISMB 100	1	95	8	250	5						150	70	6	5	2
2	ISMB 125	1	95	8	250	5						150	90	6	5	2
3	ISMB 150	1	95	8	300	5						150	100	6	5	2
4	ISMB 175	1	95	10	300	6						150	100	6	5	2
5	ISMB 200	1	125	12	300	8						150	125	6	5	2
6	ISMB 250	1	100	20	500	8						150	190	6	5	2A
7	ISMB 300	1	115	20	525	8						150	240	6	5	2A
8	ISMB 350	1	115	25	650	8						150	250	8	6	2A
9	ISMB 400	1	115	25	600	10						150	300	8	6	2A
10	ISMB 450	1	125	25	600	10						175	350	8	6	2A
11	ISMB 500	1	155	25	700	10						200	400	8	6	2A
12	ISMB 600	1	185	25	660	12	2	65.00	12	660	12	240	480	10	8	1

Note: Refer **Figure 3.21**; Material: E 250 (Fe 410 W) A; Fy = 250 MPa; Electrode – Electrode Ex40-44XX.

TABLE 5.37
Welded Splice in Channel Section – IS 800 (Working Stress)

SL No.	Member Section	Nos/ Flng	Flange Cover Plate – 1				Flange Cover Plate – 2					Web Cover Plate – 2 nos				Type
			Wide Bf₁ mm	Thick Tf mm	Length Lf mm	Weld ω₁ mm	Nos/ Flng	Wide Bf₂ mm	Thick Tf mm	Length Lf mm	Weld ω₁ mm	Wide Bw mm	Depth dw mm	Thick t_wc mm	Weld ω₂ mm	
1	ISMC 75	1	50.00	8.00	150	5.00						100	50	8.00	6.00	1
2	ISMC 100	1	60.00	8.00	200	5.00						100	70	8.00	0.19	1
3	ISMC 125	1	75.00	10.00	250	6.00						100	85	10.00	0.19	1
4	ISMC 150	1	85.00	10.00	300	6.00						100	105	10.00	0.19	1
5	ISMC 175	1	85.00	12.00	300	8.00						100	130	10.00	0.19	1
6	ISMC 200	1	85.00	12.00	300	8.00						100	155	10.00	0.25	1
7	ISMC 250	1	90.00	16.00	400	8.00						100	200	12.00	0.25	1
8	ISMC 300	1	100.00	16.00	400	8.00						120	240	12.00	0.25	1
9	ISMC 400	1	125.00	16.00	500	8.00						150	300	16.00	0.25	1

Note: Refer **Figure 3.21**; Material: E 250 (Fe 410 W) A; Fy = 250 MPa; Electrode – Electrode Ex40-44XX.

TABLE 5.38
Beams and Girder with Bearing Type Bolt – IS 800 (Working Stress) – Part I

SL No.	Beam Section	Flange Wide bf (mm)	Flange Thick tf (mm)	Web Depth d (mm)	Web Thick tw (mm)	Capacity Shear V (kN)	Capacity Moment M (kNm)	Type	Flange Cover Plate – 1 Nos	Wide Bf₁ (mm)	Thick Tf (mm)	Length Lf (mm)	Flange Cover Plate – 2 Nos	Wide Bf₁ (mm)	Thick Tf (mm)	Length Lf (mm)
1	ISMB 250	125	12.5	250	6.9	85	60	2A	1	125	16	500				
2	ISMB 300	140	12.4	300	7.5	110	80	2A	1	140	16	525				
3	ISMB 350	140	14.2	350	8.1	140	100	2A	1	140	16	550				
4	ISMB 400	140	16	400	8.9	175	140	2A	1	140	20	600				
5	ISMB 450	150	17.4	450	9.4	210	180	2A	1	150	20	675				
6	ISMB 500	180	17.2	500	10.2	250	240	2A	1	170	20	700				
7	ISMB 600	210	20.8	600	12	350	400	2	1	230	20	600				
8	PG1	600	32	1536	16	1200	4600	1	1	550.00	20	1500	2	245	20	1500
9	PG2	450	28	1250	12	750	2450	1	1	400.00	20	1000	2	175	20	1000
10	PG3	450	25	1200	12	720	2100	1	1	400.00	16	1000	2	175	16	1000
11	PG4	450	20	1150	12	690	1500	1	1	400.00	16	1000	2	175	16	1000
12	PG5	400	20	1000	12	600	1200	1	1	350.00	16	900	2	150	16	900
13	PG6	400	20	1000	12	600	1200	1	1	350.00	16	900	2	150	16	900
14	PG7	350	20	950	12	570	1000	1	1	300.00	16	900	2	125	16	900
15	PG8	300	16	900	12	540	650	1	1	250.00	12	900	2	100	12	900
16	PG9	300	16	850	10	425	600	1	1	250.00	12	850	2	100	12	850
17	PG10	250	16	750	10	375	450	1	1	250.00	10	850	2	100	10	850

Note: Refer Figures 4.20 to 29; Material: E 250 (Fe 410 W) A; Fy = 250 MPa; Bolt – HT 8.8 Grade (Bearing type).

TABLE 5.39
Beams and Girder with Bearing Type Bolt – IS 800 (Working Stress) – Part II

| SL No. | Beam Section | Web Cover Plate - 3 | | | Flange Connection Bolts | | | | | Web Connection Bolts | | | |
		Wide Bw mm	Depth dw mm	Thick t_{wc} mm	Size	Rows	No. of Bolts/Row	Nos/Side N1	Spcg. mm	Size	Group	Nos/Side N2
1	ISMB 250	250	175	6	M16	2	3	6	96	M16	4G	4
2	ISMB 300	250	210	6	M16	2	3	6	102	M16	4G	4
3	ISMB 350	250	245	8	M16	2	3	6	109	M16	6G	6
4	ISMB 400	250	280	8	M16	2	4	8	81	M16	6G	6
5	ISMB 450	275	350	8	M20	2	3	6	137	M20	6G	6
6	ISMB 500	275	400	8	M20	2	3	6	143	M20	6G	6
7	ISMB 600	275	500	8	M20	2	4	8	79	M20	8G	8
8	PG1	320	1200	12	M24	2	6	12	132	M24	16G	16
9	PG2	320	900	10	M24	2	6	12	82	M24	16G	16
10	PG3	320	900	10	M24	2	6	12	82	M24	16G	16
11	PG4	275	850	10	M20	2	6	12	87	M20	16G	16
12	PG5	275	850	10	M20	2	6	12	77	M20	16G	16
13	PG6	275	850	8	M20	2	6	12	77	M20	16G	16
14	PG7	275	750	8	M20	2	6	12	77	M20	12G	12
15	PG8	275	750	8	M20	2	5	10	97	M20	12G	12
16	PG9	275	650	8	M20	2	5	10	90	M20	12G	12
17	PG10	245	600	8	M16	2	5	10	92	M16	10G	10

Note: Refer **Figures 4.20 to 29**; Material: E 250 (Fe 410 W) A; Fy = 250 MPa; Bolt – HT 8.8 Grade.

TABLE 5.40
Beams and Girder with Friction Grip Bolt – IS 800 (Limit State) – Part I

SL No.	Beam Section	Flange Wide bf (mm)	Flange Thick tf (mm)	Web Depth d (mm)	Web Thick tw (mm)	Capacity Shear V (kN)	Capacity Moment M (kNm)	Capacity Type	Flange Cover Plate – 1 Nos	Wide Bf₁ (mm)	Thick Tf (mm)	Length Lf (mm)	Flange Cover Plate – 2 Nos	Wide Bf₁ (mm)	Thick Tf (mm)	Length Lf (mm)
1	ISMB 250	125	12.5	250	6.9	110	85	2A	1	125	16	650				
2	ISMB 300	140	12.4	300	7.5	145	115	2A	1	140	16	700				
3	ISMB 350	140	14.2	350	8.1	180	150	2A	1	140	16	850				
4	ISMB 400	140	16	400	8.9	230	195	2A	1	140	20	900				
5	ISMB 450	150	17.4	450	9.4	275	255	2A	1	150	20	950				
6	ISMB 500	180	17.2	500	10.2	335	340	1	1	170	20	700	2	60	20	700
7	ISMB 600	210	20.8	600	12	470	575	1	1	230	20	750	2	90	20	750
8	PG1	600	32	1536	16	1600	6800	1	1	550	20	1500	2	245	20	1500
9	PG2	450	28	1250	12	980	3650	1	1	400	20	1000	2	175	20	1000
10	PG3	450	25	1200	12	945	3130	1	1	400	16	900	2	175	16	900
11	PG4	450	20	1150	12	900	2390	1	1	400	16	900	2	175	16	900
12	PG5	400	20	1000	12	780	1850	1	1	350	16	700	2	150	16	700
13	PG6	400	20	1000	12	780	1850	1	1	350	16	700	2	150	16	700
14	PG7	350	20	950	12	740	1540	1	1	300	16	700	2	125	16	700
15	PG8	300	16	900	12	700	1000	1	1	250	12	500	2	100	12	500
16	PG9	300	16	850	10	550	945	1	1	250	12	450	2	100	12	450
17	PG10	250	16	750	10	490	695	1	1	250	10	800	2	100	10	800

Note: Refer **Figures 4.20 to 29**; Material: E 250 (Fe 410 W) A; Fy = 250 MPa; Bolt – HT 8.8 Grade.

TABLE 5.41

Beams and Girder with Friction Grip Bolt – IS 800 (Limit State) – Part II

SL No.	Beam Section	Web Cover Plate - 3			Flange Connection Bolts					Web Connection Bolts		
		Wide Bw	Depth dw	Thick t_{wc}	Size	Rows	No. of Bolts/Row	Nos/Side N1	Spcg.	Size	Group	Nos/Side N2
		mm	mm	mm					mm			
1	ISMB 250	250	175	6	M20	2	5	10	65	M20	4G	4
2	ISMB 300	250	210	6	M20	2	6	12	57	M20	4G	4
3	ISMB 350	250	245	8	M20	2	7	14	60	M20	4G	4
4	ISMB 400	250	280	8	M20	2	8	16	55	M20	6G	6
5	ISMB 450	275	350	8	M20	2	9	18	51	M20	6G	6
6	ISMB 500	275	400	8	M20	2	5	10	72	M20	6G	6
7	ISMB 600	275	500	8	M24	2	5	10	72	M24	8G	8
8	PG1	320	1200	12	M27	4	10	40	72	M27	16G	16
9	PG2	320	900	10	M27	4	6	24	80	M27	12G	12
10	PG3	320	900	10	M27	4	5	20	87	M27	12G	12
11	PG4	275	850	10	M24	4	6	24	72	M24	12G	12
12	PG5	275	850	10	M24	4	5	20	66	M24	10G	10
13	PG6	275	850	8	M24	4	5	20	66	M24	10G	10
14	PG7	275	750	8	M24	4	5	20	66	M24	10G	10
15	PG8	275	750	8	M24	4	3	12	81	M24	10G	10
16	PG9	275	650	8	M24	4	3	12	69	M24	8G	8
17	PG10	245	600	8	M24	2	5	10	78	M24	6G	6

Note: Refer **Figures 4.20 to 29**; Material: E 250 (Fe 410 W) A; Fy = 250 MPa; Bolt – HT 8.8 Grade.

TABLE 5.42
Channel Section with Bearing Type Bolt – IS 800 (Working Stress) – Part I

SL No.	Channel Section	Flange Cover Plate				Flange Bolt N1				Web Cover Plate			Web Bolts N2		
		Nos per Flange	Wide Bf_1	Thick Tf	Length Lf	Bolt Size	Nos per Side	Spcg s	Nos	Wide Bw	Depth dw	Thick t_{wc}	Bolt size	Nos per Side	Spcg n
			mm	mm	mm	mm		mm		mm	mm	mm	mm		mm
1	ISMC 200	1	75.00	12.00	450	M16	3	167	1	200	160	10.00	M16	3	71
2	ISMC 250	1	80.00	16.00	500	M20	3	186	1	200	200	10.00	M20	3	93
3	ISMC 300	1	90.00	16.00	600	M20	3	236	1	200	255	10.00	M20	3	118
4	ISMC 400	1	125.00	16.00	600	M20	3	236	1	200	375	12.00	M24	3	156

Note: Refer **Figure 4.33**; Material: E 250 (Fe 410 W) A; $Fy = 250$ MPa; Bolt – HT 8.8 Grade bearing type.

TABLE 5.43

Channel Section with Bearing Type Bolt – IS 800 (Working Stress) – Part II

SL No.	Channel Section	Flange		Web		Capacity of Splice Joint	
		wide	thick	depth	thick	Moment	Shear
		bf	tf	d	tw	M	V
		mm	mm	mm	mm	kNm	kN
1	ISMC 200	75	11.4	200	6.1	30	85
2	ISMC 250	80	14.1	250	7.1	40	85
3	ISMC 300	90	13.6	300	7.6	55	110
4	ISMC 400	100	15.3	400	8.6	90	170

TABLE 5.44
Beams and Girder with Slip Critical Bolt – AISC (LRFD/ASD) – Part I

SL No.	Beam Section	Flange		Web		Capacity			Flange Cover Plate – 1				Flange Cover Plate – 2			
		Wide bf	Thick tf	Depth d	Thick tw	Shear V	Moment M	Type	Nos	Wide Bf₁	Thick Tf	Length Lf	Nos	Wide Bf₁	Thick Tf	Length Lf
		inch	inch	inch	inch	kips	kip-ft			inch	inch	inch		inch	inch	inch
1	W10×17	4.01	0.33	10.11	0.24			2	1	5.00	5/16	10				
2	W12×26	6.49	0.38	12.22	0.23			2A	1	5.00	5/8	15				
3	W14×30	6.73	0.385	13.84	0.27			2A	1	6.00	5/8	15				
4	W16×36	6.985	0.43	15.86	0.295		Up to full tensile strength of flange	2A	1	6.00	5/8	18				
5	W18×40	6.015	0.525	17.9	0.315	50% of web shear		2A	1	5.00	3/4	20				
6	W21×50	6.53	0.535	20.83	0.38			2A	1	5.50	3/4	20				
7	W24×84	9.02	0.77	24.1	0.47			1	1	8.00	5/8	30	2	2.77	5/8	30
8	PG1	23.62	1 1/4	60.47	5/8			1	1	23.00	1	72	2	10.00	1	72
9	PG2	17.72	1	49.21	5/8			1	1	16.00	3/4	48	2	7.00	3/4	48
10	PG3	17.72	1	47.24	5/8			1	1	16.00	3/4	48	2	7.00	3/4	48
11	PG4	17.72	3/4	45.28	5/8			1	1	16.00	5/8	48	2	7.00	5/8	48
12	PG5	15.75	3/4	39.37	5/8			1	1	14.00	5/8	36	2	6.00	5/8	36
13	PG6	15.75	3/4	39.37	1/2			1	1	14.00	5/8	44	2	6.00	5/8	44
14	PG7	13.78	3/4	37.40	1/2			1	1	12.00	5/8	44	2	5.00	5/8	44
15	PG8	11.81	5/8	35.43	1/2			1	1	10.00	5/8	30	2	4.00	5/8	30
16	PG9	11.81	5/8	33.46	3/8			1	1	10.00	5/8	27	2	4.00	5/8	27
17	PG10	9.84	5/8	29.53	3/8			1	1	8.00	5/8	24	2	3.00	5/8	24

Notes: Refer **Figures 4.20 to 29**; Material: ASTM A36; Fy = 36 ksi; Bolt – HT 8.8 Grade.

This chart is applicable for LRFD and ASD load combinations.

Design strength of the joint: Moment – full strength; Shear – 50% of web shear capacity.

All bolts are High tensile 8.8 grade and Slip critical connection.

TABLE 5.45
Beams and Girder with Slip Critical Bolt – AISC (LRFD/ASD) - Part II

SL No.	Beam Section	Web Cover Plate - 3			Flange Connection Bolts					Web Connection Bolts		
		Wide Bw inch	Depth dw inch	Thick t_{wc} inch	Size	Rows	No of Bolts/Row	Nos/Side N1	Spcg. mm	Size	Group	Nos/Side N2
1	W10×17	8	8	0.25	M16	2	2	4	69	M16	6G	6
2	W12×26	8	10	0.25	M16	2	3	6	66	M16	6G	6
3	W14×30	8	11	0.25	M16	2	3	6	66	M16	6G	6
4	W16×36	10	13	0.25	M20	2	3	6	82	M20	8G	8
5	W18×40	10	14	0.25	M20	2	3	6	95	M20	8G	8
6	W21×50	10	17	0.31	M20	2	3	6	95	M20	8G	8
7	W24×84	10	19	0.38	M20	2	5	10	79	M20	10G	10
8	PG1	13	48	1/2	M27	2	12	24	74	M27	16G	16
9	PG2	12	39	1/2	M24	2	8	16	75	M24	16G	16
10	PG3	12	38	1/2	M24	2	8	16	75	M24	12G	12
11	PG4	12	36	1/2	M24	2	7	14	87	M24	12G	12
12	PG5	12	31	1/2	M24	2	6	12	74	M24	12G	12
13	PG6	10	31	3/8	M20	2	8	16	71	M20	12G	12
14	PG7	10	30	3/8	M20	2	8	16	71	M20	12G	12
15	PG8	10	28	3/8	M20	2	5	10	79	M20	12G	12
16	PG9	10	27	5/16	M20	2	5	10	70	M20	12G	12
17	PG10	10	24	5/16	M20	2	4	8	80	M20	10G	10

Notes: Refer **Figures 4.20 to 29**; Material: ASTM A36; Fy = 36 ksi; Bolt – HT 8.8 Grade.
This chart is applicable for LRFD and ASD load combinations.
Design strength of the joint: Moment – full strength; Shear – 50% of web shear capacity.
All bolts are High tensile 8.8 grade and Slip critical connection.

TABLE 5.46
Columns with Friction Grip Bolt – IS 800 (Limit State/Working) - Part I

SL No.	Beam Section	Flange Wide bf (mm)	Flange Thick tf (mm)	Web Depth d (mm)	Web Thick tw (mm)	Capacity Shear V (kN)	Capacity Moment M (kNm)	Capacity Type	Flange Cover Plate – 1 Nos	Wide Bf₁ (mm)	Thick Tf (mm)	Length Lf (mm)	Flange Cover Plate – 2 Nos	Wide Bf₁ (mm)	Thick Tf (mm)	Length Lf (mm)
1	ISMB 250	125	12.5	250	6.9			2A	1	125	16	700				
2	ISMB 300	140	12.4	300	7.5			2A	1	140	16	800				
3	ISMB 350	140	14.2	350	8.1			2A	1	140	16	800				
4	ISMB 400	140	16	400	8.9			2A	1	140	20	1000				
5	ISMB 450	150	17.4	450	9.4	2.5% of axial load of column section	Up to full strength of column flange (tension/compression)	2A	1	150	20	1100				
6	ISMB 500	180	17.2	500	10.2			1	1	180	20	1400	2	65.00	20	1400
7	ISMB 600	210	20.8	600	12			1	1	210	16	750	2	80	16	750
8	C1	600	32	936	16			1	1	600.00	20	1100	2	270	20	1100
9	C2	500	28	944	16			1	1	500.00	16	1050	2	220	16	1050
10	C3	450	25	950	16			1	1	450.00	16	1050	2	195	16	1050
11	C4	450	20	860	12			1	1	450.00	12	950	2	200	12	950
12	C5	400	25	650	12			1	1	400.00	16	750	2	175	16	750
13	C6	400	25	650	12			1	1	400.00	16	750	2	175	16	750
14	C7	300	20	560	12			1	1	300.00	12	790	2	125	12	790
15	C8	250	16	568	12			1	1	250.00	12	700	2	100	12	700
16	C9	250	16	468	12			1	1	250.00	12	600	2	100	12	600
17	C10	200	16	368	12			1	1	200.00	12	500	2	75	12	500

Notes: Refer **Figures 4.36; 4.21 to 29**; Material: E 250 (Fe 410 W) A; Fy = 250 MPa; Bolt – HT 8.8 Grade.
This chart is applicable for load combinations as per limit state and working stress.
Design strength: Flange force – full strength (tension / compression); Web – full compression; Shear – 2.5% of compressive strength of column.
All bolts are High tensile 8.8 grade and Friction grip connection.

TABLE 5.47
Columns with Friction Grip Bolt – IS 800 (Limit State) – Part II

SL No.	Beam Section	Web Cover Plate - 3			Flange Connection Bolts					Web Connection Bolts		
		Wide Bw mm	Depth dw mm	Thick t_{wc} mm	Size	Rows	No of Bolts/Row	Nos/Side N1	Sprg. mm	Size	Group	Nos/Side N2
1	ISMB 250	240	200	6	M20	2	5	10	78	M20	4G	4
2	ISMB 300	240	245	6	M20	2	5	10	90	M20	6G	6
3	ISMB 350	240	285	8	M20	2	6	12	72	M20	8G	8
4	ISMB 400	240	330	8	M20	2	7	14	77	M20	8G	8
5	ISMB 450	240	380	8	M20	2	8	16	73	M20	8G	8
6	ISMB 500	240	420	8	M20	2	5	10	165	M20	12G	12
7	ISMB 600	310	490	8	M24	2	5	10	82	M24	12G	12
8	C1	510	770	12	M27	4	7	28	75	M27	30g	30
9	C2	510	770	12	M27	4	6	24	85	M27	30g	30
10	C3	510	780	12	M27	4	5	20	106	M27	24g	24
11	C4	450	700	12	M24	4	5	20	97	M24	24g	24
12	C5	450	530	10	M24	4	5	20	72	M24	24g	24
13	C6	450	530	10	M24	4	5	20	72	M24	24g	24
14	C7	500	450	10	M24	2	6	12	61	M24	15g	15
15	C8	450	450	10	M24	2	5	10	66	M24	15g	15
16	C9	450	370	10	M24	2	4	8	71	M24	13g	12
17	C10	450	275	10	M24	2	3	6	81	M24	9g	9

Notes: Refer **Figures 4.36; 4.21 to 29**; Material: E 250 (Fe 410 W) A; Fy = 250 MPa; Bolt – HT 8.8 Grade.
This chart is applicable for load combinations as per limit state and working stress.
Design strength: Flange force – full strength (tension / compression); Web – full compression; Shear – 2.5% of compressive strength of column.
All bolts are High tensile 8.8 grades and Friction grip connection.

TABLE 5.48
Columns with Slip Critical Bolt – AISC (LRFD/ASD) – Part I

SL No.	Beam Section	Flange Wide bf (inch)	Flange Thick tf (inch)	Web Depth d (inch)	Web Thick tw (inch)	Capacity Shear V (kips)	Capacity Mom M (kip-ft)	FCP-1 Type	FCP-1 Nos	FCP-1 Wide Bf₁ (inch)	FCP-1 Thick Tf (inch)	FCP-1 Length Lf (inch)	FCP-2 Nos	FCP-2 Wide Bf₁ (inch)	FCP-2 Thick Tf (inch)	FCP-2 Length Lf (inch)
1	W10×17	4.01	0.33	10.11	0.24			2	1	5.00	5/16	20				
2	W12×26	6.49	0.38	12.22	0.23			2A	1	5.00	5/8	22				
3	W14×30	6.73	0.39	13.84	0.27			2A	1	6.00	1/2	22				
4	W16×36	6.99	0.43	15.86	0.30			2A	1	6.00	5/8	24				
5	W18×40	6.02	0.53	17.90	0.32			2A	1	5.00	3/4	24				
6	W21×50	6.53	0.54	20.83	0.38			2A	1	5.50	3/4	28				
7	W24×84	9.02	0.77	24.10	0.47	30% of web shear capacity	Up to 40% of flange force (tension or compression)	1	1	8.00	5/8	28	2	2.77	5/8	28
8	Col C1	23.62	1.26	36.85	0.63			1	1	23.00	3/4	40	2	10.00	0.75	40
9	Col C2	19.69	1.13	37.17	0.63			1	1	18.00	3/4	40	2	8.00	3/4	40
10	Col C3	17.72	0.98	37.40	0.63			1	1	16.00	5/8	40	2	7.00	5/8	40
11	Col C4	17.72	0.79	33.86	0.63			1	1	16.00	1/2	36	2	7.00	1/2	36
12	Col C5	15.75	0.98	29.53	0.63			1	1	14.00	3/4	30	2	6.00	3/4	30
13	Col C6	15.75	0.79	27.95	0.47			1	1	14.00	5/8	30	2	6.00	5/8	30
14	Col C7	13.78	0.79	25.98	0.47			1	1	12.00	5/8	40	2	5.00	5/8	40
15	Col C8	11.81	0.63	20.39	0.47			1	1	10.00	1/2	30	2	4.00	1/2	30
16	Col C9	11.81	0.63	18.43	0.47			1	1	10.00	1/2	30	2	4.00	1/2	30
17	Col C10	9.84	0.63	16.46	0.47			1	1	8.00	1/2	25	2	3.00	1/2	25

Notes: Refer **Figures 4.40; 4.21 to 29**; Material: ASTM A36; Fy = 36 ksi; Bolt – ASTM A490M

This chart is applicable for load combinations as per LRFD and ASD.

Design strength: The ends of column shafts are machined for bearing. It is assumed that 60% compression will be transferred by bearing at contact face; Shear – 30% of web shear capacity.

All bolts are ASTM A490 M and Slip critical connection.

TABLE 5.49
Columns with Slip Critical Bolt – AISC (LRFD/ASD) - Part II

SL No.	Beam Section	Web Cover Plate - 3			Flange Connection Bolts					Web Connection Bolts		
		Wide Bw inch	Depth dw inch	Thick t_{wc} inch	Size	Rows	No of Bolts/Row	Nos/Side N1	Spcg. inch	Size	Group	Nos/Side N2
1	W10×17	8	8	0.25	M16	2	4	8	65	M16	4G	4
2	W12×26	10	10	0.25	M20	2	4	8	72	M20	6G	6
3	W14×30	10	11	0.25	M20	2	4	8	72	M20	6G	6
4	W16×36	10	13	0.25	M20	2	4	8	80	M20	6G	6
5	W18×40	10	14	0.25	M20	2	4	8	80	M20	6G	6
6	W21×50	10	17	0.31	M20	2	5	10	73	M20	8G	8
7	W24×84	10	20	0.31	M20	2	5	10	73	M20	12G	12
8	Col C1	13	29	1/2	M27	4	5	20	102	M27	16G	16
9	Col C2	12	30	1/2	M24	4	5	20	105	M24	20G	20
10	Col C3	12	30	1/2	M24	4	4	16	140	M24	20G	20
11	Col C4	12	27	1/2	M24	4	3	12	185	M24	20G	20
12	Col C5	12	24	1/2	M24	4	4	16	98	M24	16G	16
13	Col C6	10	22	3/8	M20	4	4	16	106	M20	16G	16
14	Col C7	10	21	3/8	M20	2	7	14	74	M20	16G	16
15	Col C8	10	16	3/8	M20	2	5	10	79	M20	12G	12
16	Col C9	10	15	5/16	M20	2	5	10	79	M20	10G	10
17	Col C10	10	13	5/16	M20	2	4	8	85	M20	10G	10

Notes: Refer **Figures 4.40; 4.21 to 29**; Material: ASTM A36; Fy = 36 ksi; Bolt – ASTM A490M.

This chart is applicable for load combinations as per LRFD and ASD.

Design strength: The ends of column shafts are machined for bearing. It is assumed that 60% compression will be transferred by bearing at contact face; Shear – 30% of web shear capacity.

All bolts are ASTM A490 M and Slip critical connection.

REFERENCES

1. Steel Construction Manual AISC – Fourteenth edition.
2. IS 800: 2007 General Construction in Steel – Code of Practice.
3. AISC 360 – 10 Specification for Structural Steel Buildings.

6 Plates Showing Typical Joints of Industrial Structures

6.1 INTRODUCTION

This chapter exhibits some plates covering actual style of drawing details that should be used in structural steel design drawing. The reader will know about the method of preparation of drawing and dimensioning of joints.

Following are the list of items covered in this chapter.

a) Secondary beam to main beam connection – field welded.
b) Secondary beam to main beam connection – bolted joint.
c) Column splice – shop welded.
d) Column splice – bolted joint.
e) Splice of rolled and plated section – shop welded.
f) Splice of rolled and plated section – bolted.
g) Column bracing joints – sheet 1.
h) Column bracing joints – sheet 2.
i) Floor bracing joint.
j) Sliding joint.
k) Crane girder joints – sheet 1.
l) Crane girder joints – sheet 2.
m) Column base plate.
n) Cage ladder – front entry.
o) Cage ladder – side stepped.
p) Staircase joints – sheet 1.
q) Staircase joints – sheet 2.
r) Grating stairs.
s) Handrails.
t) Grating and checkered plate floor.
u) Roof purlin and side girts joints.
v) Embedded plate – sheet 1.
w) Embedded plate – sheet 2.
x) Embedded plate – sheet 3.

DOI: 10.1201/9781003539124-6

6.2   PLATES SHOWING TYPICAL JOINTS OF INDUSTRIAL STRUCTURES

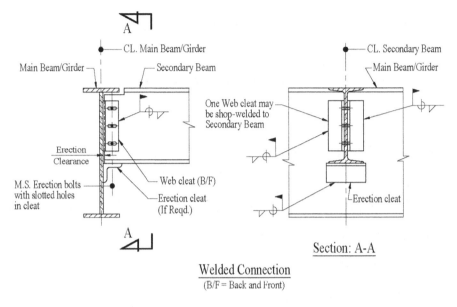

FIGURE 6.1   Secondary beam to main beam connection – field welded.

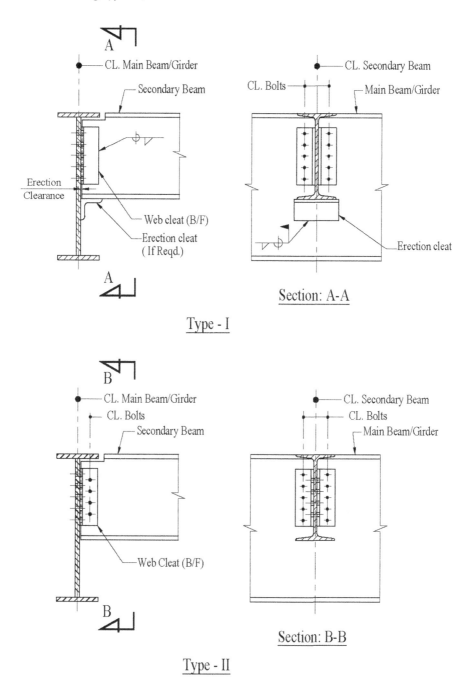

A

CL. Main Beam/Girder

Secondary Beam

CL. Secondary Beam

CL. Bolts

Main Beam/Girder

Erection
Clearance

Web cleat (B/F)

Erection cleat
(If Reqd.)

A

Erection cleat

Section: A-A

Type - I

B

CL. Main Beam/Girder

CL. Bolts

Secondary Beam

CL. Secondary Beam

CL. Bolts

Main Beam/Girder

Web Cleat (B/F)

B

Section: B-B

Type - II

FIGURE 6.2 Secondary beam to main beam connection – bolted joint.

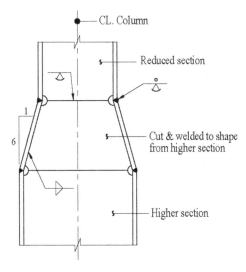

Unequal Section

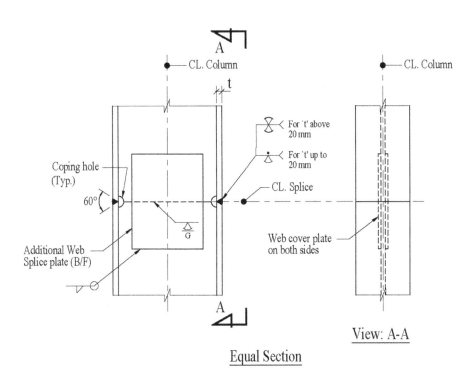

Equal Section

View: A-A

FIGURE 6.3 Column splice – shop welded.

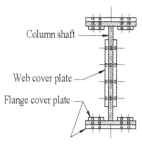

Section: B-B

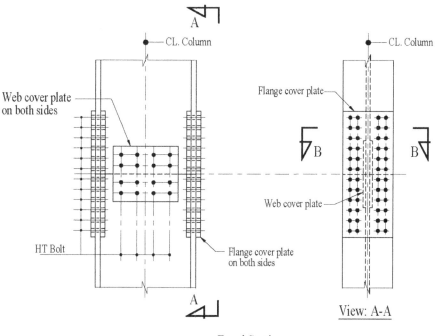

Equal Section   View: A-A

FIGURE 6.4   Column splice – bolted joint.

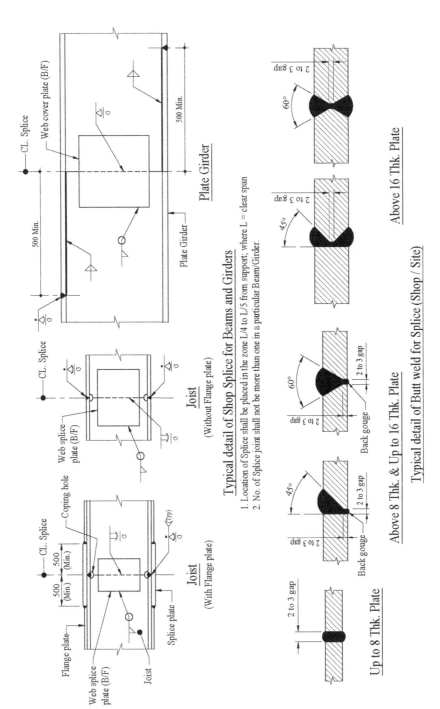

FIGURE 6.5   Splice of rolled and plated section – shop welded.

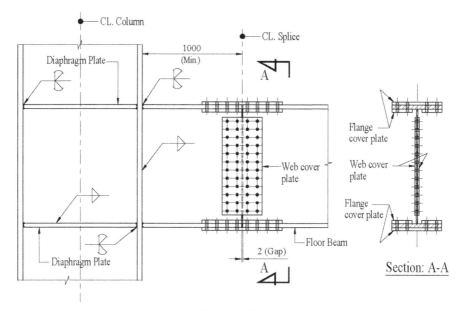

Plated Section

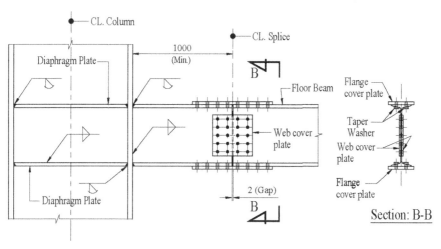

Rolled Section

FIGURE 6.6   Splice of rolled and plated section – bolted.

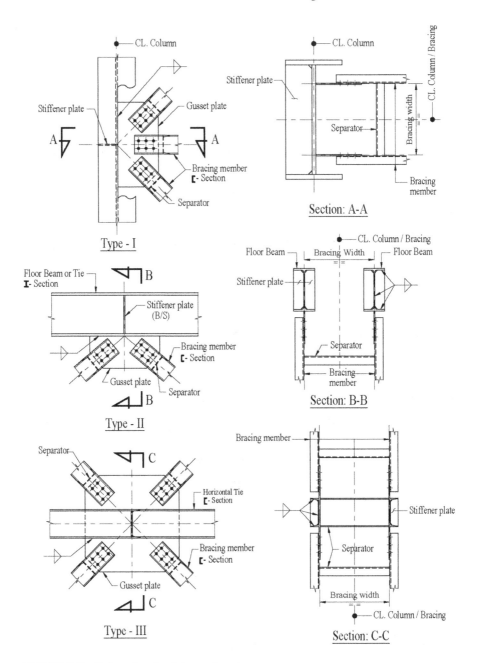

FIGURE 6.7   Column bracing joints – sheet 1.

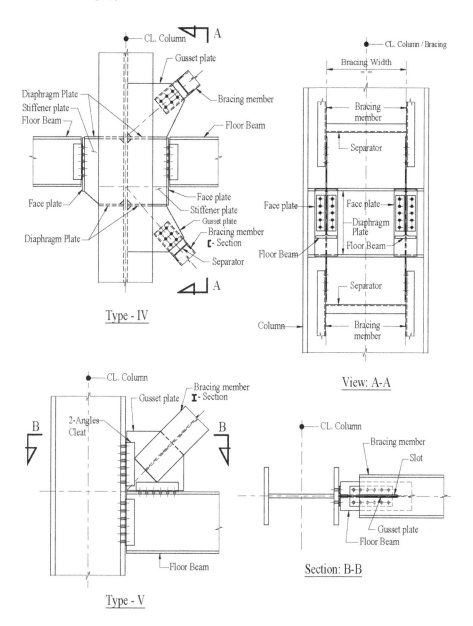

FIGURE 6.8 Column bracing joints – sheet 2.

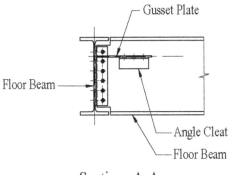

Section: A-A

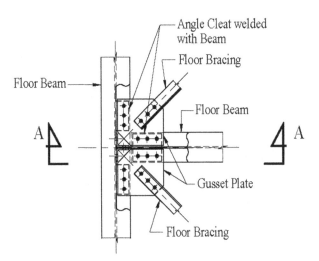

FIGURE 6.9   Floor bracing joint.

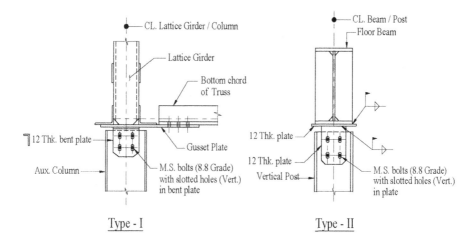

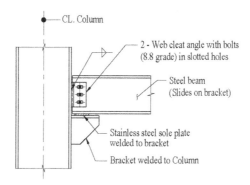

FIGURE 6.10   Sliding joint.

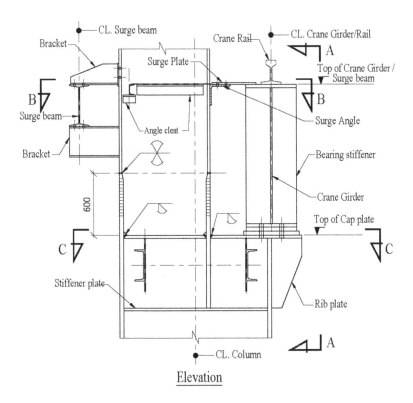

Elevation

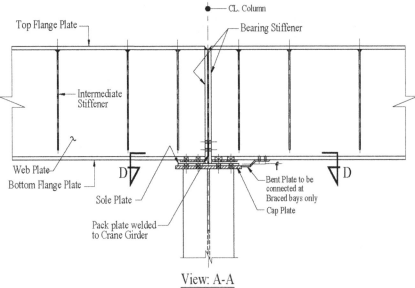

View: A-A

FIGURE 6.11   Crane girder joints – sheet 1.

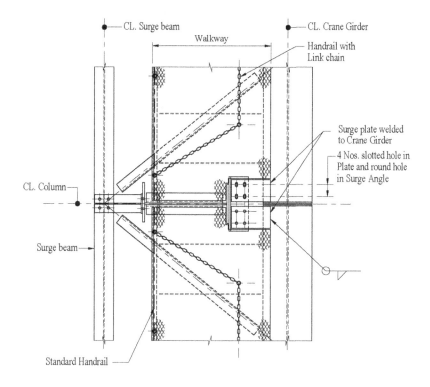

Section: B-B

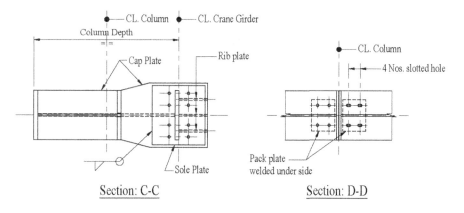

Section: C-C   Section: D-D

FIGURE 6.12   Crane girder joints – sheet 2.

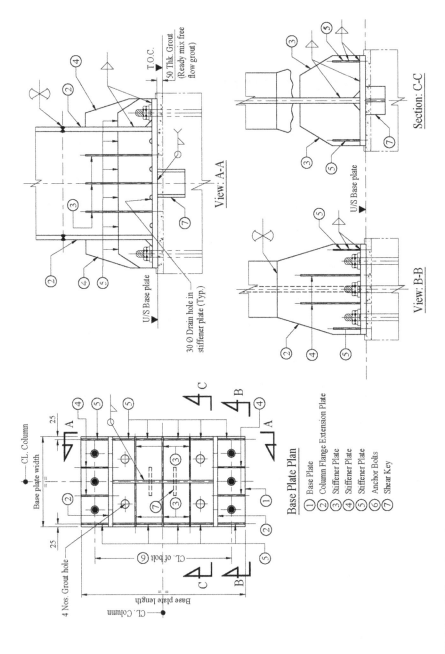

FIGURE 6.13   Column base plate.

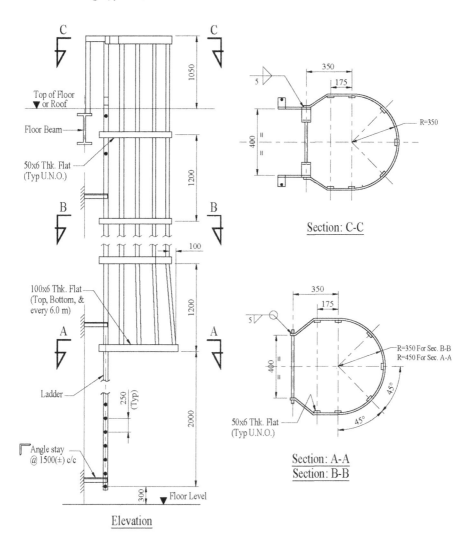

FIGURE 6.14   Cage ladder – front entry.

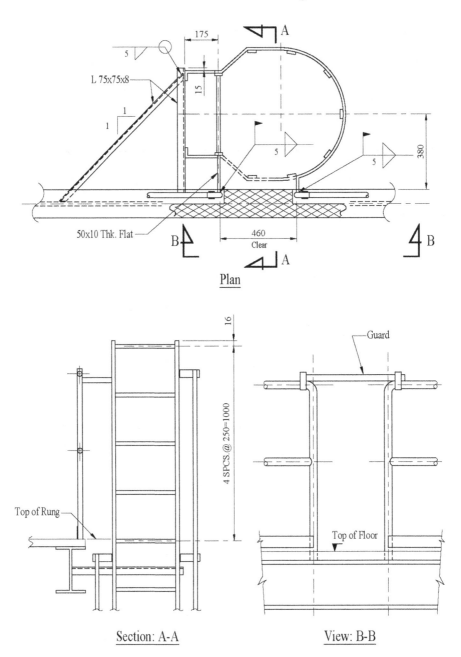

FIGURE 6.15   Cage ladder – side stepped.

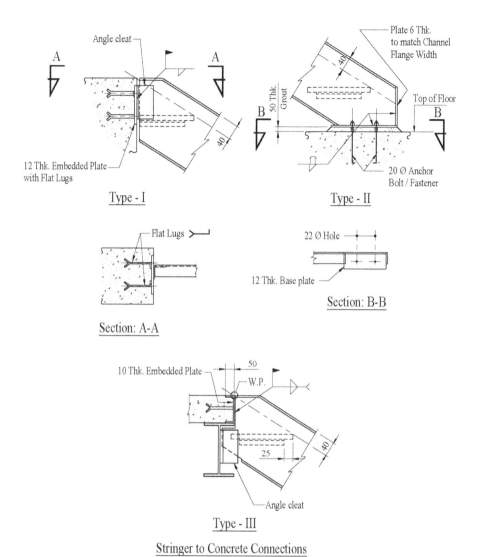

FIGURE 6.16 Staircase joints – sheet 1.

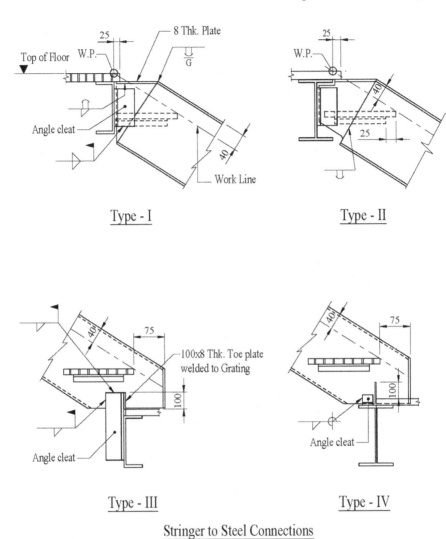

Type - I

Type - II

Type - III

Type - IV

Stringer to Steel Connections

FIGURE 6.17   Staircase joints – sheet 2.

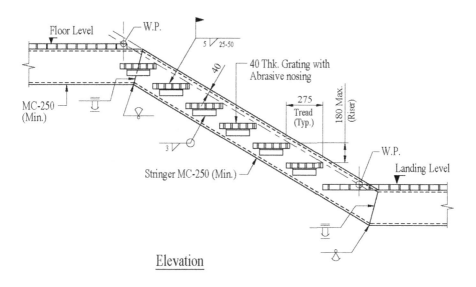

Elevation

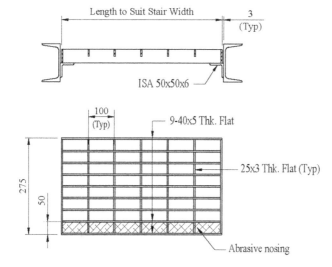

Detail of Grating Tread

1. Pattern shall be according to Standard Electroforged Grating system
2. All Treads should be Galvanized

FIGURE 6.18   Grating Stairs.

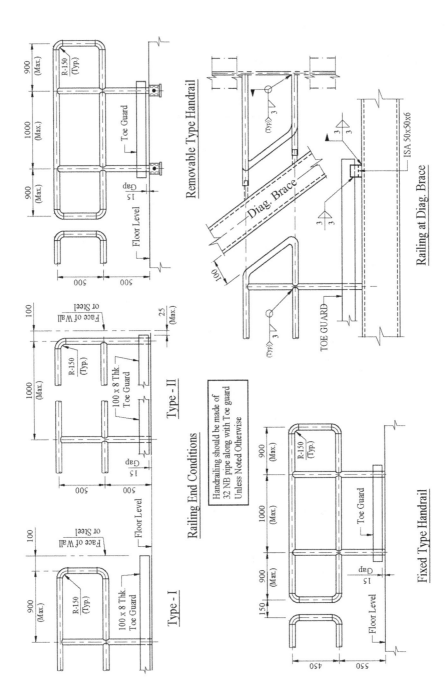

FIGURE 6.19 Handrails.

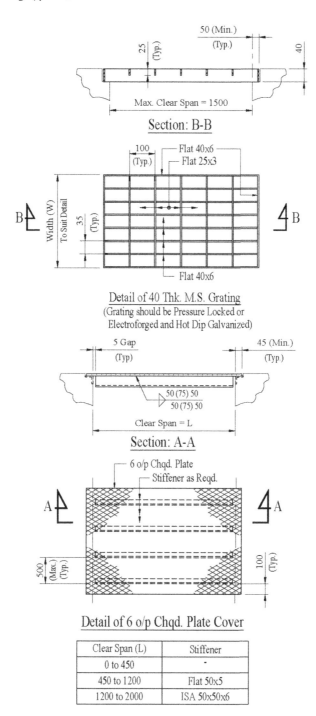

FIGURE 6.20 Grating and checkered plate floor.

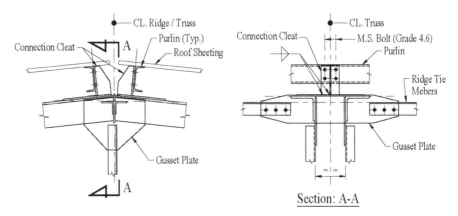

CL. Ridge / Truss

Connection Cleat

Purlin (Typ.)

Roof Sheeting

A

Gusset Plate

A

Roof Purlin Support

CL. Truss

Connection Cleat

M.S. Bolt (Grade 4.6)

Purlin

Ridge Tie Mebers

Gusset Plate

Section: A-A

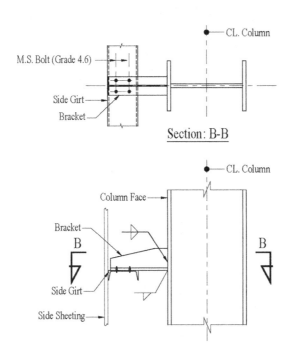

CL. Column

M.S. Bolt (Grade 4.6)

Side Girt

Bracket

Section: B-B

CL. Column

Column Face

Bracket

B

Side Girt

Side Sheeting

B

FIGURE 6.21   Roof purlin and side girts joints.

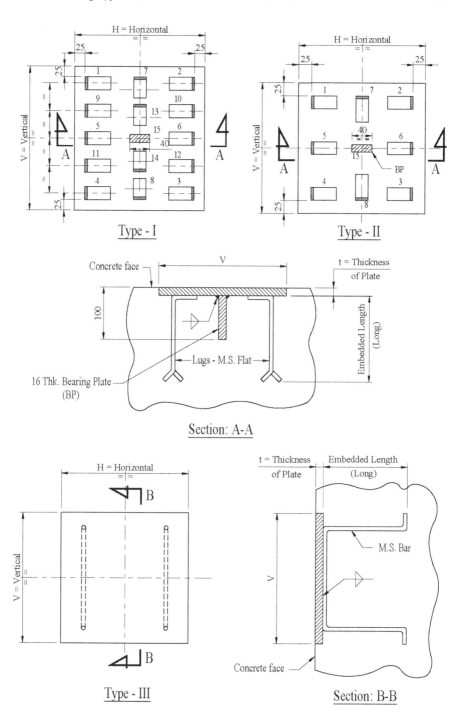

FIGURE 6.22   Embedded plate – sheet 1.

Rectangular Plates

Item Marked	'V'x'H'	't'	Lugs Type - Nos.	Location	Bearing Plate (BP)
EP-201	200 x 100	8	Round 12 - 6 Nos.	1 Thru. 6	Nil
EP-302	300 x 200	10	Flat 50 - 4 Nos.	1 Thru. 4 & 15	100 x 40 x 16 Thk.
EP-403	400 x 300	12	Flat 50 - 10 Nos.	1 Thru. 8, 13, 14	Nil
EP-504	500 x 400	12	Flat 50 - 14 Nos.	1 Thru. 14	100 x 40 x 16 Thk.
EP-604	600 x 400	16	Flat 50 - 14 Nos.		
EP-251	250 x 150	12	Flat 25 - 6 Nos.	1 Thru. 6	Nil
EP-252	250 x 200	12	Flat 50 - 6 Nos.	1 Thru. 4 & 15	100 x 40 x 16 Thk.
EP-352	350 x 250	12	Flat 50 - 6 Nos.	1 Thru. 6 & 15	
EP-452	450 x 250	12	Flat 50 - 10 Nos.	1 Thru. 6, 9 Thru. 12, & 15	
EP-652	650 x 250	16	Flat 50 - 14 Nos.		

Square Plates

Item Marked	'V'='H'	't'	Lugs Type - Nos.	Location	Bearing Plate (BP)
EP-100	100	8	Round 12 - 4 Nos.		Nil
EP-150	150	8	Round 12 - 4 Nos.		
EP-200	200	10	Flat 25 - 4 Nos.	1, 2, 3, 4	Nil
EP-250	250	12	Flat 25 - 8 Nos.	1 Thru. 8 & 15	100x40x 16 Thk.
EP-300	300	12	Flat 50 - 8 Nos.		
EP-400	400	12	Flat 50 - 10 Nos.	1, 2, 3, 4 5, 6, 7, 8, 13, 14	Nil
EP-450	450	12	Flat 50 - 10 Nos.		
EP-500	500	12	Flat 50 - 14 Nos.	1 Thru. 14	Nil
EP-600	600	16	Flat 50 - 14 Nos.		

FIGURE 6.23   Embedded plate – sheet 2.

Lugs for Embedded Plate

Type	Description
Flat - 25	25x6 Thk.x300 Lg.
Flat - 50	50x6 Thk.x450 Lg.
Round - 12	12Ø x200 Lg.
Round - 16	16Ø x300 Lg.

Corner Angle

Corner Angle Mark	Section Used	Lug
ISA - 45	ISA 45x45x5	10 Dia. Bar
ISA - 50	ISA 50x50x6	
ISA - 75	ISA 75x75x6	25x6 Thk. Flat
ISA - 100	ISA 100x100x8	

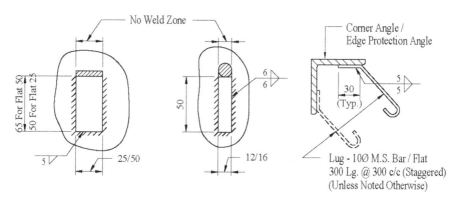

Welding of Lugs

FIGURE 6.24 Embedded plate – sheet 3.

REFERENCES

1. Steel Construction Manual AISC – Fourteenth edition.
2. IS 800: 2007 General Constructions in Steel – Code of Practice.

Index

Pages in *italics* refer to figures and pages in **bold** refer to tables.

Printed in the United States
by Baker & Taylor Publisher Services